MAKING SENSE

OF

OSHA

COMPLIANCE

JEFFREY W. VINCOLI, CSP

Government Institutes, Inc., 4 Research Place, Rockville, Maryland 20850, USA.

01 00 99 98 97 5 4 3 2 1

Library of Congress Cataloging-in-Publication Data

Vincoli, Jeffrey W.
Making sense of OSHA compliance / by Jeffrey W. Vincoli.
p. cm.
Includes bibliographical references and index.
ISBN: 0-86587-535-9
1. Industrial safety--Law and legislation--United States. 2. Industrial hygiene--Law and legislation--United States.
I. Title.
KF3570.V56 1997
344.73' 0465--dc21
96-11924
CIP

Printed in the United States of America

At a time when the future of OSHA is in question, this book is dedicated to the representatives of future American productivity:

To my young nieces and nephews

Heather Louise Parker
Nikki-Lynn Vincoli
Karri Ann Parker
Robert J. Parker
Eric R. Gordon
Christie Lynn Bowling
Erica Jane Fryou

and to all the other members of tomorrow's work force.

May their places of employment remain free from the hazards that can threaten their safety and health.

TABLE OF CONTENTS

CHAPTER 2—THE OCCUPATIONAL SAFETY AND HEALTH ACT OF 1970 15

CHAPTER 3—OSHA STANDARDS, THE RULEMAKING PROCESS, AND ENFORCEMENT 35

Preface

This text was developed for use by the safety and health professional as well as those persons responsible for safety and health issues in the workplace. The complexity of OSHA regulations and the critical nature of compliance requires some basic knowledge of U.S. Administrative Law in general and the Occupational Safety and Health Act in particular. But since most safety and health practitioners are not lawyers, much of the theory of law has been replaced by real-world examples of OSHA compliance based upon the author's fifteen years of experience in this arena.

Making Sense of OSHA Compliance should therefore be used primarily as a tool for persons having management responsibility for occupational safety and health. It is a practical guide, with practical application of the theories of safety and health law and common sense suggestions for application to the real-world problems of OSHA compliance. While a great deal of information is presented in this text, it is not possible to include all that there is to know concerning OSHA compliance in a limited volume such as this. However, readers who desire a fundamental understanding of the Occupational Safety and Health Act of 1970 and the agencies created under it should find satisfaction in the materials provided.

Because it is anticipated that this text will be used primarily as a reference guide, those reading through these materials may find some degree of repetition from chapter to chapter and from subject area to subject area. This is intentional, to ensure proper emphasis on key aspects of safety and health law as they relate to a particular facet of the compliance process. It is also partially unavoidable since the Act itself tends to have some degree of redundancy as it moves from section to section.

In short, managers, supervisors, engineers, executives, compliance officers, quality inspectors, technicians, union representatives, and, of course, safety and health personnel and trainers, should benefit from the information contained in this text.

Jeffrey W. Vincoli, CSP

About the Author

Jeffrey W. Vincoli is president and principal consultant of J.W. Vincoli & Associates, a safety, health, and environmental training and consulting firm in Titusville, FL. He is a former manager of safety and environmental engineering for McDonnell Douglas Aerospace's Kennedy Space Center Division, with over 14 years experience in space, missile and strategic defense programs. An award-winning author of six books on a range of EH&S topics, Vincoli has had numerous articles published in *Professional Safety* and other safety publications. He makes frequent presentations at professional development courses and safety meeting nationwide. Vincoli is a Certified Safety Professional (Board of Certified Safety Professionals), Registered Environmental Professional (National Registry of Environmental Professionals), Certified Safety Specialist (World Safety Organization), Certified Hazard Control Manager (International Hazard Control Manager Certification Board), and Certified Accident Investigator-Aerospace (DOE, NRC, NASA). A member of the American Society of Safety Engineers, National Environmental Health Association, System Safety Society, and the editorial board of *Occupational Hazards*, Vincoli is also a member of the ASSE Professional Development Conference Planning Committee and currently chairs ASSE's PDC Speaker Selection Subcommittee. As part of People to People International's American People Ambassador Program, he has led delegations of U.S. safety and health professionals visiting the People's Republic of China, Hong Kong, Vietnam, Australia, and New Zealand, and was a delegate on a visit to the former Soviet Union. Vincoli holds a master of science degree in aeronautical science, *magna cum laude*, and a master of business administration degree in aviation management, *magna cum laude*, both from Embry-Riddle Areonautical University.

Acknowledgments

The author wishes to thank the following individuals and organizations who have assisted in the development of this publication:

First, George S. Brunner for his contributions and suggestions and for faithfully proofreading the materials presented in this text.

Second, Steven S. Phillips, Jeffrey S. Drye, and Frank Beckage for their assistance with research and for helping to ensure the overall accuracy of the information and materials presented.

Finally, a special thanks to Alexander Padro and the staff at Government Institutes for allowing me the opportunity to write this text and for ensuring the overall quality of this publication.

MAKING SENSE

OF

OSHA

COMPLIANCE

Chapter 1

ADMINISTRATIVE LAW AND OCCUPATIONAL SAFETY AND HEALTH

OVERVIEW

This chapter provides a fundamental understanding of the administrative law process and how it pertains to occupational safety and health in the United States. Understanding the basics of this process is essential for effective OSHA compliance.

The OSHAct of 1970 provides the constitutional basis for the promulgation of occupational safety and health regulations. It is referred to as the *statutory mandate* which created the Occupational Safety and Health Administration. As a federal agency, OSHA is influenced by the three branches of government.

Since it is the Congress which provides OSHA the authority to promulgate and enforce regulations, the *legislative branch* can influence and control OSHA. Congress also passes operating budgets, thereby directly controlling the activities of the agency.

The *executive branch* develops and submits the national budget and, therefore, can influence decisions made by OSHA. More directly, however, the executive branch is responsible for appointing top level officials in the federal agency. The opinions and positions of these political appointees will almost always be closely aligned with those of the President.

The *judicial branch* can also influence the activities of OSHA by adjudicating and interpreting the meaning and applicability of a particular standard, thereby setting precedents for all future similar cases.

The *administrative rulemaking* process with respect to OSHA, and the promulgation of occupational safety and health regulations, will also be discussed. While OSHA uses both *formal* and *informal rulemaking,* the agency generally operates under the less stringent (and less time-consuming) informal process. When some aspects of formal and informal rulemaking are combined (which is often the case under OSHA) the process is referred to as *hybrid rulemaking.* After issuing a *Notice of Proposed Rulemaking (NPR)* in the *Federal Register (FR),* as required under the *Administrative Procedures Act (APA),* OSHA can hold hearings, solicit opinions, establish advisory committees, and accept comment from the public before issuing a new standard. Once issued, the new regulation is formally codified in the *Code of Federal Regulations (CFR).* Throughout this process, there are numerous opportunities for citizens and corporations to get involved and influence the development of the final rule.

INTRODUCTION

To understand the force of law behind federal legislation such as the Occupational Safety and Health Act, a brief explanation of the administrative law process and the constitutional basis for safety and health regulations is necessary.

The term *administrative law* applies to that body of law that governs the methods by which administrative agencies make and implement decisions. Federal administrative law is based on specific provisions of the U.S. Constitution and in various other federal statutes. Within this regulatory framework of administrative law, occupational safety and health legislation obtains the *force of law.*

THE CONSTITUTIONAL BASIS FOR SAFETY AND HEALTH REGULATION

Figure 1-1 shows the various relationships that exist between administrative agencies, such as the Occupational Safety and Health Administration (OSHA), and the three branches of government. Under most circumstances, administrative agencies in the United States are created by the legislative branch, usually by an Act. Once an agency is established, it is managed and run by the executive branch of government. Periodically, federal agencies are subjected to review by the judicial branch. Understanding how federal agencies like OSHA operate requires knowing how each of these branches can influence and even control the agency's behavior.

Figure 1-1: The relationship between federal agencies and the three branches of government.

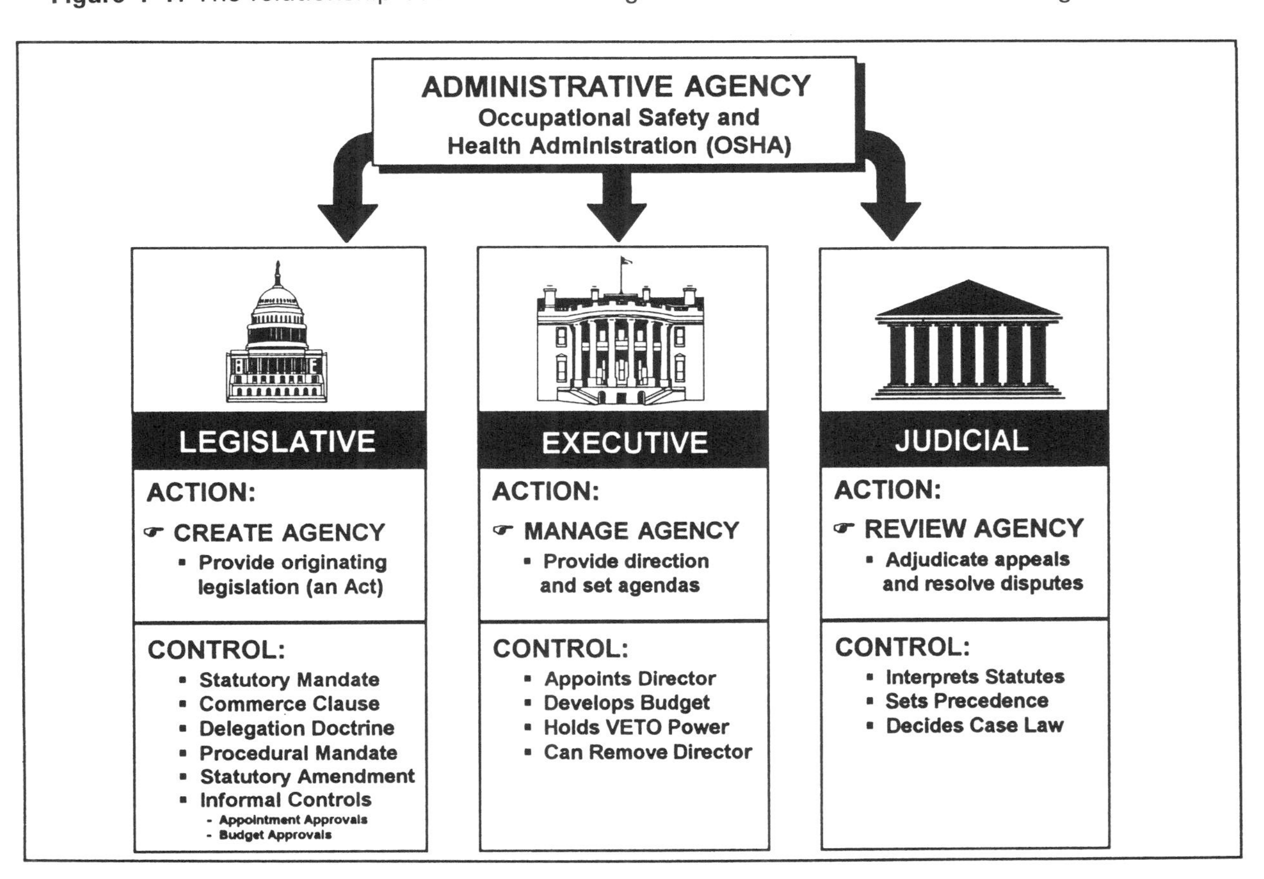

The Role of the Legislative Branch

The legislative branch of government can direct and influence the behavior of a federal agency through a series of formal and informal controlling actions. These include:

Statutory Mandates. A statutory mandate is a formal directive from Congress granting a particular agency the authority to act in a given area of concern, such as occupational safety and health. An Act of Congress begins the administrative process by either creating a new federal agency to address that area of concern or granting new authority, responsibility, and powers to an existing agency. When an Act creates a new agency, it is often referred to as the agency's *originating statute* or enabling legislation. It is this originating statute that provides the agency's statutory mandate. The Occupational Safety and Health Act of 1970 (OSHAct) was the originating statute of OSHA and it is the enabling legislation from which OSHA obtains its statutory mandate to address issues and concerns related to occupational safety and health.

More than one agency can be created by a single originating statute. The OSHAct created not only OSHA, but also the National Institute for Occupational Safety and Health (NIOSH) and the Occupational Safety and Health Review Commission (OSHRC). While OSHA is the only agency of the three charged with the actual administration and enforcement of regulatory requirements (i.e., rule-making authority), the remaining two agencies do have very specific responsibilities that can and do influence the formulation, promulgation, and adjudication of occupational safety and health law in the United States. The functions of both NIOSH and OSHRC will be explained in more detail in the next chapter. For now, suffice it to say that NIOSH is responsible for conducting research and recommending policies, procedures, and approaches to OSHA. These actions can obviously influence the direction of any new or existing OSHA regulation (primarily in the area of occupational health). OSHRC adjudicates conflicts between OSHA and cited employers and presides over appeals to ensure OSHA is properly and fairly applying its rules and regulations.

Just as a single statutory mandate can create more than one agency, more than one agency can administer the same originating statute. This can become confusing from the perspective of both rule-making and compliance. For example, both OSHA and the Environmental Protection

Agency (EPA) have the authority to regulate worker exposure to harmful chemicals. This means that persons responsible for ensuring regulatory compliance must research all appropriate agency requirements to determine applicability.

The Commerce Clause. The U.S. Constitution grants Congress the power to authorize administrative agencies such as OSHA to act. The commerce clause grants Congress the power "to regulate commerce..... among the States" (Article I, Section 8). This commerce power has been broadened considerably over the years through judicial review and interpretation. Congress now has broad powers under this clause to regulate in the general areas of health, safety, and the environment. For example, in the name of interstate commerce, Congress can promulgate statutes requiring inspection and maintenance of rail cars used to ship chemical products between states in order to ensure that interstate shipment of bulk chemicals by rail will not be interrupted due to leaking or improperly loaded containers.

The Delegation Doctrine. Most statutory mandates to agencies, especially those pertaining to safety, health, and the environment, are broad and non-specific. As a result, agencies such as OSHA and EPA are granted considerable discretion when it comes to the interpretation and enforcement of statutory requirements. In the strictest interpretation of law, this situation may be considered unconstitutional. The relevant constitutional principle, known as the *delegation doctrine,* is based on the understanding that the Congress, as the duly elected representative of the people, is the repository of all legislative power. Only the people can grant this power to the Congress.

According to the delegation doctrine, Congress can not, in turn, delegate this legislative power to another party, such as an administrative agency, because the agency has not been elected by the people. Under strict application of this doctrine, Congress is required to provide reasonably clear and specific statutory standards to guide agency decision making. In reality, the standards enacted by Congress are usually anything but clear. For example, anyone who has read the Clean Air Act Amendments or the Process Safety Management Standard can see that clarity was certainly not a consideration during their development.

Violation of the delegation doctrine has not been raised at issue since 1936. In that year, the U.S. Supreme Court struck down two separate statutes on the grounds that they granted improperly broad decision-making authority to administrative agencies. However, since that time, the number of administrative agencies has increased dramatically and agency decision-making has become the principal method of federal regulation. The courts have also relaxed the delegation doctrine considerably over the years. Today, it is common to find federal agencies operating with extremely broad authority (the Internal Revenue Service is an excellent example).

The Procedural Mandate. In addition to following the specific mandates of the originating statutes, federal agencies must also adhere to the more general procedural requirements of numerous other statutes. This is known as the *procedural mandate.* In the simplest of terms, the originating statute provides guidance to the agency on *what* decisions must be made in a particular area of concern (e.g., safety and health), while the procedural mandate provides guidance on *how* those decisions are to be made. Procedural mandates are basically rules governing the operation of an agency's decision-making process. The Administrative Procedures Act (APA) of 1946 (5 U.S.C. Section 551) is the primary means by which Congress controls the procedures of the various federal agencies. The APA establishes standard operating procedures for:

- Agency rulemaking and adjudication;
- Judicial review of administrative decision making; and
- Citizen access to the administrative and judicial processes.

A potential problem can arise when the originating statute specifies procedural requirements that may be in conflict with those established in the APA. The OSHAct, for instance, contains specifications for rulemaking that differ from those found in the APA. In general, when such conflicts arise, the more specific requirements established in the originating statute will apply. When the originating statute does not address such issues, the agency must then follow the provisions of the APA.

Interpretation of the Statutory Mandate. A primary task of a federal agency, and for the courts when reviewing agency decisions, is the *interpretation of the statutory mandate.* The broader and less specific the mandate, the more difficult it is to define the intent of Congress. To determine this intent, both the agency and the courts begin the process by examining the language of the statute itself. Where the language is sufficiently clear and unambiguous, the agency need not and, under the APA, may not pursue the matter further. It must execute the intentions of Congress as expressed in the language of the statute. However, in the majority of the cases, the intent of Congress is not clear based on the language contained in the statute and, subsequently, intent becomes a matter of interpretation.

By examining the history behind the statute, any prior drafts of the statute, reports to congressional committees with regard to the statute, and substantially all other available information about the intent of the statute, the courts can attempt to interpret the intended meaning of the statute. The courts will also examine how the agency has interpreted the statute in the past. Unless the agency's precedents conflict with the legislative history of the statute, the courts will generally defer to the agency's interpretation. This is why formally contesting an OSHA citation based solely upon conflicts in the interpretation of a statute is rarely successful.

Statutory Amendments and Informal Controls. When Congress believes that an agency is not acting in accordance with legislative intent or Congressional directives, it can take a number of actions. The most direct method is for Congress to *amend* a particular statute to clarify its mandate to the agency. In theory, this approach is the surest and most direct way to remedy the problem. In reality, it typically takes too long, so Congress often uses other less direct and informal means to influence an agency's decision-making process. For instance, members of Congress can issue very public "statements" that criticize the actions of an agency or express Congressional dissatisfaction with the agency. Such statements can sometimes cause the agency to react favorably to Congressional pressures. Congress can also use committee hearings to question, warn, and even chastise agency leaders who refuse to run their agency in concert with Congressional intent.

Perhaps the most effective control Congress can use to influence an agency's actions is the annual budget. Agency officials must come before Congress to explain and justify the annual funding proposal for their agency. This provides Congress a powerful opportunity to influence controls over the agency by suggesting an increase or threatening a decrease in overall funding.

The Role of the Executive Branch

While most administrative agencies are created by the legislative branch (EPA, an exception, was created by Presidential Executive Order in 1970), administrative agencies reside within the executive branch of government. Therefore, the executive branch can execute considerable controlling influence over the actions of the administrative agency. The primary means of influence is the President's control of the appointment process. Most statutes that create an agency also permit the President to appoint the agency's top decision-making officials. These "political appointments" are subject to the approval of the Senate. By selecting candidates for these key administrative positions, the President can effectively, albeit indirectly, control the direction of the agency for at least a four year period.

The power to *appoint* a person to office is accompanied by the power to *remove* a person from office. The intent is to allow each new administration the freedom (within the bounds of applicable statutory mandates) to influence the course of the agencies that operate under its direction. However, inevitable conflicts arise because the course often favored by the new administration frequently differs from that which is favored by the Congress. The Senate approval process can, therefore, take very long and can result in a disapproval of the President's choice. While the President still has the option to appoint another candidate, the Senate must again undertake the long, arduous, and expensive approval process before the candidate can be accepted.

As with Congress, the executive branch can also execute considerable control over an agency during the budget process. Although final approval of the national budget requires Congressional action, the budget is first developed and proposed by the chief executive. Hence, the President can directly influence the actions of an agency by providing for particular programs in the national budget or, conversely, restricting funding in a particular area. Also, since 1980, the President has required the Office of Management and Budget (OMB) to oversee an economic analysis of all proposed major regulations. This means that the cost of any major regulation proposed by OSHA must be evaluated before the President will budget the funds required to implement the regulation. This can significantly delay the activities and actions of OSHA.

The Role of the Judicial Branch

In the absence of any statutory or constitutional amendment, the ultimate arbitrators of the meaning of a particular statute or constitutional provision are the courts. This means that a court can invalidate an Act of Congress if it finds the statutory language to be in violation of the Constitution. Hence, what an agency can or must do is what the courts say it can or must do. Congress can amend a statute to avoid a judicial interpretation that it does not favor, but even the new statutory language will eventually face potential scrutiny by a reviewing court.

The courts have tremendous power over the final actions of an agency. However, it must be clearly understood that the courts (including the U.S. Supreme Court) are absolutely powerless to act until someone brings a lawsuit to their attention. Even if the courts believe an agency action or decision is unconstitutional or in violation of the agency's statutory mandate, they *can not* initiate any action until a lawsuit has been filed. With the Supreme Court, even after a lawsuit has been filed, the high court can not act until the case has progressed upward from the lowest court, which could take several years.

WHAT IS ADMINISTRATIVE RULEMAKING?

In general terms, an agency executes its statutory mandate by promulgating administrative regulations. These regulations are first published in the Federal Register (FR), usually dated the same day as the regulation was promulgated. They are later codified more permanently in the U.S. *Code of Federal Regulations (CFR).* Unless the regulations are challenged successfully either in the courts or by Congress, they carry the force of law.

Federal agencies perform a variety of functions within the bounds of their statutory mandate. Some functions may be clearly *legislative,* others may be *judicial,* and still others may be defined as *enforcement.* When an agency engages in rulemaking it is primarily involved in the legislative aspects of its mandate. In effect, rulemaking creates agency policy as well as regulations. For instance, when OSHA establishes a specific workplace standard, it has really done two things. First, it has engaged in the rulemaking process to create a regulation of law and, second, it has established a requirement (i.e., a policy) to which all employers who are covered by the OSHAct must adhere. But OSHA does more than just develop and promulgate administrative rules, it also enforces them.

Many regulatory statutes also create some type of administrative tribunal to adjudicate disputes that often arise when the agency uses its enforcement power. As mentioned previously in this Chapter, the OSHAct created OSHRC for just this purpose. Employers cited for allegedly violating the Act or some regulation promulgated under the Act can appeal to OSHRC. The Commission hears the facts and determines whether a violation of the Act has occurred. The OSHRC does not issue broad policy statements, but provides an interpretation of the law based upon the facts presented. In doing so, the OSHRC has engaged in *adjudication,* not rulemaking, which is an important distinction.

OSHA's Formal Rulemaking Process

When OSHA wishes to promulgate a new rule, the APA requires them to first publish a General Notice of Proposed Rulemaking (or NPR) in the Federal Register. The NPR must provide a statement of the time, place, and nature of the public rulemaking proceedings. It must provide reference to the legal authority under which the rule is proposed. And it must provide a description of either the terms of the proposed rule or the subjects and issues involved. Once the notice has appeared in the FR, the APA prescribes two general procedures OSHA can use in promulgating the new administrative regulations: *formal rulemaking* and *informal rulemaking.* The former can be quite cumbersome in process and procedure. Formal rulemaking is therefore only performed if it is specifically required by Congress in the originating statute. In the absence of such requirements, informal rulemaking is used.

Following the notice in the FR, informal rulemaking, often referred to as *notice and comment rulemaking,* requires that OSHA provide "interested parties an opportunity to participate in the rulemaking through submission of written data, views, or arguments with or without opportunity for oral presentation." Unlike formal rulemaking, it does not require a hearing, although OSHA may hold one if it so desires. And unlike formal rulemaking, it allows the agency to look beyond any hearing records in making rules. Another significant difference between formal and informal rulemaking pertains to evidence requirements. When courts review OSHA's actions under informal rulemaking, OSHA is not held to the "substantial evidence" test required under formal proceedings. Rather, the agency must only prove that their decisions and determinations are not "arbitrary" or "capricious" (See Figure 1-2).

Informal Rulemaking by OSHA

The statutory mandate for OSHA does not specify the use of formal rulemaking procedures. Hence, rulemaking by OSHA is conducted according to the informal notice and comment process. No formal trial-type proceedings are required. There are, however, some notable differences in this process with regard to OSHA. Congress provided additional opportunities for public input in the OSHAct (beyond that allowed for in the APA). This mixture of formal and informal rulemaking procedures is known as *hybrid rulemaking.*

The OSHAct provides for an informal public hearing as part of the rulemaking process. When promulgating a standard, OSHA must hold a public hearing if requested by "any interested person," which means that any employer, organization, or person who expresses an interest in the proposed rule can participate. In this way, people can and do become part of the rulemaking process. Even though OSHA is not required to provide for the cross-examination of witnesses at such a hearing (a requirement under formal rulemaking), it will usually do so anyway as a matter of policy.

OSHA is also authorized to submit certain regulatory issues to *advisory committees* which consist of persons from outside the agency who may have certain expertise in a given area. These committees do not supersede OSHA's regulatory powers or responsibilities. They only provide input on specific technical and policy issues that arise in the course of agency activities (See Figure 1-2).

THE ROLE OF CITIZEN AND/OR CORPORATE INVOLVEMENT IN RULEMAKING

Like most statutory mandates the OSHAct contains provisions which allow the affected public considerable opportunity to influence agency rulemaking. As discussed previously, the NPR provides the opportunity for comment *prior* to the promulgation of any substantive regulation. In reality, few people read or pay much attention to the daily Federal Register and even fewer actually provide comment. However, those that do are taken quite seriously and their comments are viewed as representative of the broad general population. In the area of occupational safety and health, it is typical to find that those who do respond to an NPR will be labor unions, trade organizations, public interest groups, and representatives of the affected industries (usually employers or their agents). They often participate in the process and, therefore, represent the views and positions of their constituencies in the rulemaking process.

There are also other methods that allow additional opportunities for citizen and corporate involvement in administrative proceedings.

Initiation of Rulemaking

The APA requires every agency, including OSHA, to give an interested party the right to file a petition requesting a rule be issued, amended, or repealed. Although this stipulation does not actually require the agency to take the action(s) requested, it does provide an avenue for the interested public to become involved. Furthermore, unless the petition is blatantly frivolous, the agency may also be required to provide a statement of its rationale if it decides not to act on the request.

Access to Agency Proceedings/Records

Nearly all OSHA activity is a matter of public record and, therefore, is generally accessible to the public. Three statutes in particular, which were added to the APA in the 1970s, are quite useful in this regard.

The Government in Sunshine Act (5 U.S.C., Section 552b). With certain exceptions, requires that "every portion of every meeting of an agency shall be open to public observation." The exceptions are designed primarily for the protection of personal privacy, trade secrets, agency enforcement efforts, and internal agency personnel rules and practices.

The Freedom of Information Act, or "FOIA" (5 U.S.C., Section 552). Requires each agency to make all of its records "promptly available to any person" upon request, subject to a similar set of exceptions as those noted above. Under FOIA, the agency must respond to any request within ten working days. Anyone who is denied access to records to which they are entitled may take the agency to federal district court to secure access and may be entitled to recover attorney's fees and the costs of the suit.

Figure 1-2: The various steps in the rulemaking process.

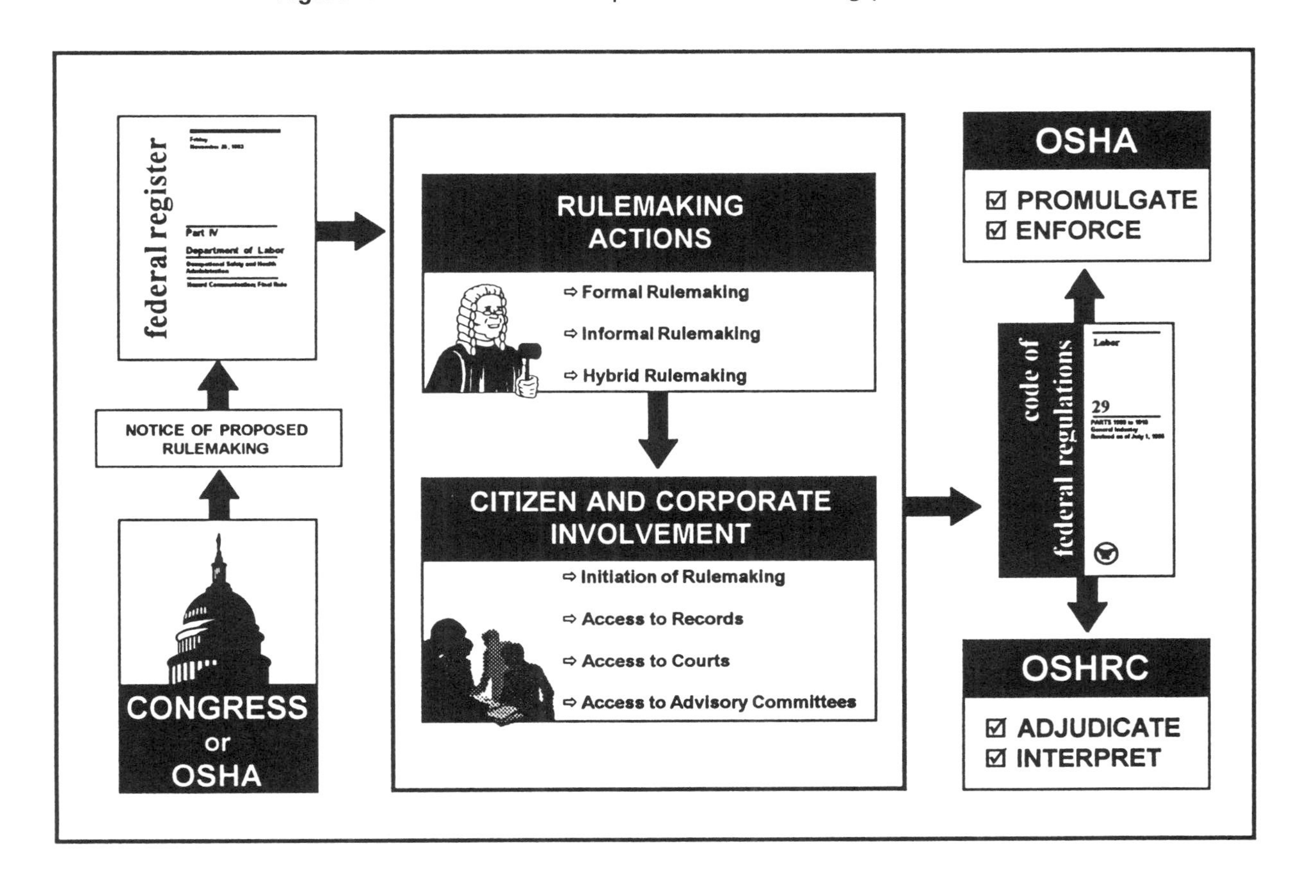

The Privacy Act (5 U.S.C., Section 552a). Grants broad access (with certain specific exceptions) to agency records pertaining to oneself. The Act provides a right to correct inaccurate references to oneself in agency records and limits an agency's disclosure of any personal information.

Access to Advisory Committees

The proceedings and records of any OSHA advisory committee are generally open to the public. The Federal Advisory Committee Act, or "FACA" (5 U.S.C., Appendix I) requires that, with certain exceptions, the meetings of advisory committees be open to the general public and the records of these committees be made available to the general public. FACA also provides the public with some control over the composition of the advisory committee. It requires that the membership of the committee be "fairly balanced" so as to represent each point of view to be presented and the functions to be performed. This provision was designed to prevent the committees from being unduly biased on either side of an issue. Hence, employers should look to the provisions of FACA as a useful tool when seeking to have input to OSHA decision making.

Access to the Courts

The public also has limited access to the courts to seek judicial review of OSHA performance. The OSHAct contains provisions which specify direct access by the public to the courts for review of certain agency decisions. In reality, most employers and other interested parties will seldom seek judicial review of OSHA's performance. It is time-consuming and can be very expensive. However, if ample reason exists, such avenues are available to employers.

Chapter 2

THE OCCUPATIONAL SAFETY AND HEALTH ACT OF 1970

OVERVIEW

This chapter will introduce the Occupational Safety and Health Act (OSHAct) with a prelude of the historical developments which led to the enactment of this legislation.

From ancient times, worker safety and health was generally recognized as an issue of importance. However, it was not until the dawn of the twentieth century that any concerted effort towards legislating workplace safety occurred in the United States. Various states attempted to regulate the issue within their borders, but without any serious success. By 1970, the concern for worker health and safety had reached such a level of awareness amongst the American population that the Congress promulgated the OSHAct. The Act created three federal agencies with different responsibilities. The Occupational Safety and Health Administration (OSHA), residing within the Department of Labor, was charged with *promulgation* and *enforcement* of regulations and standards. The National Institute for Occupational Safety and Health (NIOSH) was created within what is now the Department of Health and Human Services as a non-regulatory body with the responsibility to research health and safety issues and *recommend* standard practices. To ensure unbiased *adjudication of disputes* that may arise under the Act, the Occupational Safety and Health Review Commission (OSHRC) was established.

Key provisions of the Act itself will also be discussed, focusing on the authority granted to the three agencies created under the Act. Understanding the roles and responsibilities of each agency is a key element in effective OSHA compliance. Finally, because compliance is not entirely possible without an understanding of employer and employee responsibilities under the Act, a brief discussion on the expected actions of both employers and employees will conclude this chapter.

INTRODUCTION

Occupational illness, disease, and injury are not developments of modern technology. Nor were they new to the pioneers of the Industrial Revolution. Their existence has been acknowledged and even documented since ancient times. As far back as the Babylonian Empire (circa 2100 B.C.), for example, the *Code of Hammurabi* prescribed punishment for those who were found to be at fault in an accident resulting in the injury or death of another worker. Of course, such actions did nothing to address the elimination of workplace hazards. However, the ancient Code does provide evidence of mankind's early recognition and understanding that occupational safety and health must somehow be preserved.

General awareness of the significance of workplace hazards, injuries, disease, and death did not reach any appreciable level in the United States until the late 1800s. During this period, America was in the midst of its first serious effort at industrial expansion. Most of the early occupational safety and health initiatives at this point in history were enacted at the state level. For instance, in 1877, Massachusetts promulgated the very first law protecting and ensuring the safety of workers. Over the next two decades, numerous states followed with similar pieces of legislation and, by 1900, most heavily industrialized states had put into place some form of law requiring employers to ensure the control of certain workplace hazards. It must be mentioned, however, that most of these early state programs were extremely weak, poorly funded, improperly staffed, and highly fragmented in approach, enforcement, and effectiveness.

It was not until the early years of the twentieth century before any serious legislation for worker safety appeared in the United States. This chapter provides an introduction to the history of this legislation which led to the promulgation of the Occupational Safety and Health Act of 1970. A discussion on the Act itself will follow to provide the reader with the information needed to understand the Act, including employer and employer responsibilities, and to ensure compliance with its intentions.

The History of Federal Safety and Health Legislation

Early in this century, federal regulation in the United States addressing safety and health issues focused primarily on very specific occupations that were considered to be "dangerous" (such as merchant seaman, railroad workers, and miners). There were other, related efforts occurring at the same time which centered on safety research. For instance, the federal Office of Industrial Hygiene and Sanitation, under the U.S. Public Health Service, and the U.S. Bureau of Labor Standards examined instances of lead poisoning in industry, radium poisoning in the watch industry, hazards in brass foundries, phosphorus-induced disease, and the dangers of the dusty trades. However, even with all of this activity occurring in the scientific and technical communities, broad-based regulation was, by all accounts, very slow in coming.

Recalling the discussion from Chapter 1 about public (citizen) involvement in the modern-day rulemaking process, it is of particular interest to note that the first real federal initiatives in safety and health regulation transpired as a result of public pressure on their Congressional representatives. Public interest in occupational safety and health grew steadily during the early decades of the 1900s as a result of an increasing number of workplace accidents. People who experienced private and personal tragedies caused by occupational injury, disease, and death also began to make their concerns known to Congress. It took this level of public involvement to prompt the federal government to finally take the problem of workplace safety and health more seriously.

In 1934 the Bureau of Labor Standards was created as the first permanent federal agency with the specific task to promote safety and health in America's work force. Two years later, the Walsh-Healey Public Contracts Act initiated the first federal regulatory role in occupational safety by directing the Department of Labor to ensure that all federal contractors met minimum safety and health standards.

The 1950s and 1960s saw a steady increase in workplace safety and health issues in both the private and public sectors and, subsequently, on Capital Hill. Federal involvement in setting standards intensified and, in 1969, Congress passed the Federal Coal Mine Health and Safety Act. Although this Act was industry-specific, it established the Congressional mind-set for the broader legislation to come the following year.

THE OCCUPATIONAL SAFETY AND HEALTH ACT

By the end of the 1960s, the National Safety Council estimated that more than 14,000 deaths and two million injuries were occurring annually in American workplaces. At the same time, the environmental movement in the United States was also gaining speed. The general public was growing more and more concerned over the nation's responsibility to protect human health and the environment from toxic substances contained in the air and water supply. This country's first-ever "Earth Day," which was held in April of 1970, indicates the level such concern had achieved. One spin-off from all these activities was an increasing awareness of the workplace environment, where the potential and actual exposure to toxic substances was even greater.

In 1968, President Johnson proposed and submitted to Congress a bill to establish America's first comprehensive occupational safety and health program. The bill received heavy opposition from industry and it was dead before ever reaching the floor of Congress. But, public concern continued to press for some form of government control on workplace safety and health. The call for government action was further pursued during the Nixon Administration. By 1970, it was clear to Congress that federal action truly was required to reduce the growing number of work-related injuries, illnesses, and fatalities. In addition, the thousands of new chemicals that were entering interstate commerce each year (recall the discussion of the *commerce clause* from Chapter 1) compounded Congressional leaders' concerns about known occupational hazards, such as those associated with asbestos workers, they had failed to control. Congress' challenge was to determine how to control workplace safety and health and what the Department of Labor's role would be.

After much compromise and conflict, the 91st Congress finally enacted the Occupational Safety and Health Act (29 U.S.C. Section 51), which was signed into law on 29 December 1970. As briefly described in Chapter 1, the Act created and placed the Occupational Safety and Health Administration (OSHA) within the Department of Labor. It created and placed the National Institute for Occupational Safety and Health (NIOSH) within the Department of Health and Human Services (formerly the Department of Health, Education, and Welfare). It also created the Occupational Safety and Health Review Commission (OSHRC) as an independent agency (see Figure 2-1).

Figure 2-1: The three agencies created under the Occupational Safety and Health Act of 1970.

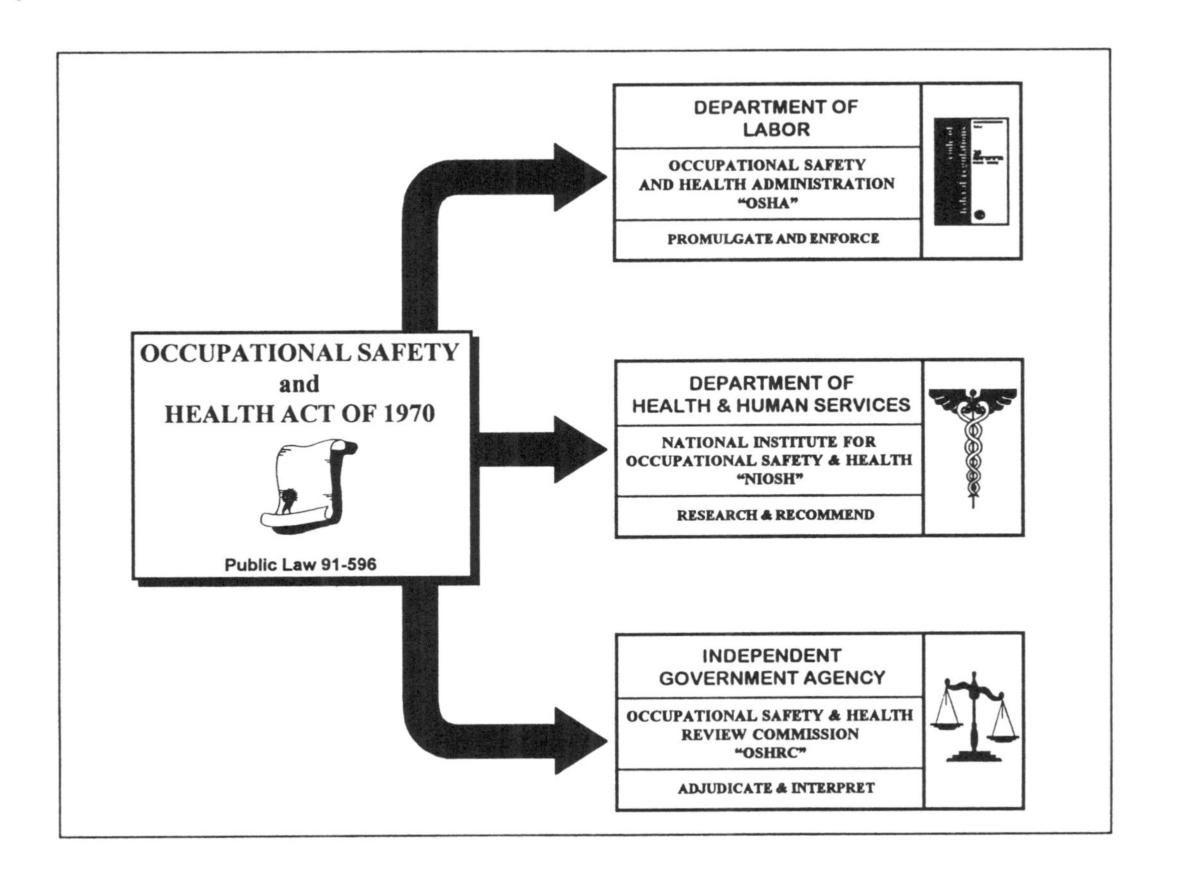

Table 2-1: Sample of OSHA Standards and the original consensus or reference standards from which they were derived either in whole or in part.

OSHA SECTION	STANDARD TITLE	CONSENSUS SOURCE OR REFERENCE	ORIGINAL TITLE OF CONSENSUS STANDARD
1910.30	Other Working Surfaces	CS 202-56 (1961)	Industrial Lifts and Hinged Loading Ramps
1910.67	Vehicle-Mounted Elevating and Rotating Work Platforms	AWS B3.0-41	Standard Qualification Procedure
1910.103	Hydrogen	ASME	Boiler and Pressure Vessel Code, Section VIII, 1968
1910.106	Flammable and Combustible Liquids	NFPA 30-1969	Flammable and Combustible Liquids Code
1910.111	Storage and Handling of Anhydrous Ammonia	ASTM A 395-68	Ductile Iron for Use at Elevated Temperatures
1910.134	Respiratory Protection	ANSI Z88.2-1969	Standard Practice for Respiratory Protection
1910.141	Sanitation	ANSI Z4.1-1968	Minimum Requirements for Sanitation in Place of Employment
1910.145(f)	Specifications for Accident Prevention Signs and Tags	ANSI Z35.2-1968	Specifications for Accident Prevention Tags
1910.157	Portable Fire Extinguishers	ANSI/NFPA 10 CGA C-6	Portable Fire Extinguishers Standards for Visual Inspection of Compressed Gas Cylinders
1910.177	Servicing Multi-Piece and Single Piece Rim Wheels	NFPA 231-1970	General Indoor Storage
1910.179	Overhead and Gantry Cranes	CMAA 61	Specification for Overhead Traveling Cranes
1910.180	Crawler Locomotive and Truck Cranes	SAE 765 (1965)	Recommended Practice: Crane Loading Stability Test Code
1910.252	Welding, Cutting, and Brazing	API 2201 (1963)	Welding or Hot Tapping on Equipment Containing Flammable
1910.261	Pulp, Paper, and Paperboard Mills	IME Pamphlet 17, 1960	Safety in the Handling and Use of Explosives
1910.307	Hazardous (classified) Locations	NFPA70C-74	Hazardous Locations Classifications

ACRONYM	ORGANIZATION NAME
ANSI:	American National Standards Institute
API:	American Petroleum Institute
ASME:	American Society of Mechanical Engineers
ASTM:	American Society for Testing Materials
AWS:	American Welding Society
CGA	Compressed Gas Association
CMAA	Crane Manufacturer's Association of America, Inc.
CS:	Commercial Standard
IME:	Institute of Makers of Explosives
NFPA:	National Fire Protection Association
SAE:	Society of Automotive Engineers

STANDARD SETTING UNDER THE OSHACT

At the time of its enactment, Congress, the President, industry, and organized labor were in agreement that occupational safety and health standards were needed. But, creating these standards from scratch would be time consuming. So Congress included provisions in the Act, under sections 6(a) and 6(b), to require the new OSHA to adopt existing national *consensus standards* within two years of the Act's passage without engaging in further rulemaking. The American National Standards Institute (ANSI) and other private-sector organizations had worked with business, government, the scientific community and others to produce standards on a consensus basis for decades. By 1970, literally hundreds of ANSI standards existed for a broad range of topics in workplace safety and health. Table 2-1 shows examples of current OSHA standards and their original consensus standard sources. Please note that many OSHA Standards may have been derived either in whole or in part from numerous existing consensus standards.

OSHA moved quickly on this mandate. By 1971, it had adopted thousands of consensus standards making the once voluntary compliance a matter of law. Most of these standards pertained to safety issues (as opposed to health). Unfortunately, many were already outdated, overly specific, or not really applicable to workplace safety. Approximately 400 of these standards did pertain to health, including threshold limit values (TLVs) recommended by the American Conference of Governmental Industrial Hygienists (ACGIH), and exposure levels recommended by ANSI. Once adopted by OSHA, these consensus standards became OSHA's permissible exposure limits (PELs). However, they included no requirements for exposure monitoring, medical surveillance of exposed workers, recordkeeping, hazard warnings, or employee training. Also, these PELs were established at a level thought to be sufficient to protect the "average" worker and did not consider the sensitized employee or the person who might already be suffering from disease or illness.

Over the years since their adoption, many guidelines were subsequently changed or revised by the ACGIH and/or ANSI. But OSHA has not kept up the pace in updating its PELs from the pre-1970 consensus levels. Subsequently, nearly all the Section 6(a) standards remain as originally adopted by OSHA. By 1988, OSHA finally attempted to initiate changes. The agency proposed a plan to update these old standards by replacing them with current ACGIH TLV levels. However, the revised standard which actually incorporated a total of 376 updated PELs came under tremendous scrutiny from the technological community, industrial organizations, labor unions, and other interested parties. The result is typical of the standard updating process. The revised standard was still under review by the end of 1996 and the old, out-dated PEL standards remain those which must be complied with by law. It is true that many employers typically attempt to adhere to the current TLVs set by ACGIH anyway. However, in the final analysis, these TLVs do not carry the force of law and will not until OSHA adopts them and makes them the new PELs. This situation exemplifies the cumbersome process of standard setting under the OSHAct. It helps explain why new or revised standards take so very long to materialize.

Which Employers Are Covered and Exempted?

When Congress passed the OSHAct, it used its powers under the commerce clause of the Constitution to the fullest extent possible. The Act was made to apply to any employer engaged in any business that affects commerce. Employers have, from time to time, attempted to use this aspect of the law as a loophole for exemption of its requirements. By asserting that the Act does not apply to their activities since they do not "affect commerce," these employers attempted to exclude themselves from the provisions of the Act. However, the OSHRC and the appellate courts usually conclude that any activity which impacts interstate commerce in any manner affects commerce.

Case precedents for the court's conclusion were established early in the history of OSHA. One case involved an employer who was clearing land with the expectation that the acreage would be used to produce wine grapes. OSHA cited the employer for violation of the Act and the employer argued that its activities were purely intrastate (i.e., within the state) and thereby exempted from the interstate commerce decision. The appellate court rejected the argument concluding that the production of wine grapes affected interstate commerce since the wine product would most likely be sold across state lines (Goodwin v. OSHRC 1976). With the exception of state governments, this decision made it reasonable to assume that virtually any employer having employees is subject to the provisions of the OSHAct unless they are exempted because the safety and health of their employees are regulated by another agency. Having one employee is sufficient to warrant coverage under the Act, even if that one employee is the sole person employed by a corporation.

It should be noted that Section 4(b)(1) of the OSHAct provides an exemption from coverage when "other federal agencies....exercise statutory authority to prescribe or enforce standards or regulations affecting occupational safety and health." Over the years, many employers have challenged this provision but, for the most part, the requirement still stands. In every case where a challenge to OSHA applicability in place of other federal agency coverage becomes the issue, the cited employers must prove that OSHA regulations do not apply to their situation because some other federal agency has jurisdiction. In most cases, the defense will prevail if the other federal agency has regulated and enforced its regulations such that it completely addresses the same safety and/or health issue(s) OSHA is attempting to regulate.

Federal Preemption and State Plans

Some employers are subject to the jurisdiction of state Occupational Safety and Health Plans. Table 2-2 shows the states and territories that regulate the safety and health of employers under their own individual, OSHA-certified plans. Congress has determined that any state or territory who adopted and enforced an occupational safety and health program at "least as effective" as the federal program will have jurisdiction over workplace safety and health issues.

In general, states who adopt their own plans are still subject to supervision by the U.S. Secretary of Labor through OSHA. In practice, OSHA supervises or exercises concurrent jurisdiction with the states until it is determined that the state's program meets the "equally effective" criteria. At such time, OSHA certifies the state plan and withdraws federal enforcement over issues covered by the plan. It is noted that states sometimes do not assume responsibility over all employers within their jurisdiction. For example, maritime employment is still subject to federal jurisdiction is some Pacific Coast states. OSHA does retain the right to reassert authority if conditions seem to warrant such action.

Table 2-2: States having their own OSHA-certified state plans.

OSHA-APPROVED STATE PLANS	
Alaska	New Mexico
Arizona	North Carolina
California	Oregon
Colorado	Puerto Rico
Connecticut	South Carolina
Hawaii	Tennessee
Indiana	Utah
Iowa	Vermont
Kentucky	Virgin Islands
Maryland	Virginia
Michigan	Washington
Minnesota	Wyoming
Nevada	

THREE AGENCIES AND THEIR FUNCTIONS

The OSHAct created three separate agencies. Figure 2-1 portrays each of these agencies. Congress split the administrative roles between two completely independent agencies within the executive branch.

The OSHAct authorized the Department of Labor to create safety and health standards and regulations, conduct on-site compliance inspection and to prosecute employers found or alleged to be in violation of the Act. The Occupational Safety and Health Review Commission (OSHRC) was created to adjudicate disputes between OSHA and employers. The National Institute for Occupational Safety and Health (NIOSH), created under Section 22 of the OSHAct, is authorized to develop and establish recommended occupational safety and health standards, conduct research activities to understand and improve workplace safety and health conditions, and develop and provide employee education and training materials in the areas of occupational safety and health. It is important to note that NIOSH can only *recommend* new standards, improvements or revisions to existing standards, and methods of performing safety and health related activities. The Institute has no regulatory powers. When OSHA chooses to act upon a NIOSH recommendation and/or adopts a recommended standard, only then will the NIOSH recommendation carry the force of law.

The following briefly describes the role of each agency under the Act, with reference to the particular sections of the OSHAct that provide the authority for agency action. Many of these topics will be expanded and discussed further in subsequent chapters of this text and are briefly presented here as an introduction only.

OSHA's Role: Adopting Standards and Enforcing Compliance

OSHA, under Section 6(a-c) of the OSHAct, was granted the power to adopt standards and grant temporary and permanent variances from complying with the standards. If OSHA concludes that they are necessary, the agency can adopt temporary emergency standards, which are valid for six months without engaging in rulemaking procedures. While emergency standards are designed to become standards via rulemaking proceedings during those six months, they are frequently rejected by the courts.

Standards vs. Regulations

Rather than address safety and health conditions, OSHA regulations address required safety and health actions or activities. For example, a standard can determine the acceptable level of dust in the air at a workplace, while a regulation specifies recordkeeping requirements or inspection activities.

Inspection Priorities. In terms of enforcement, OSHA is authorized to enter places of business at reasonable hours to conduct unannounced inspections in a reasonable manner. In fact, OSHA has established priorities for workplace inspections, as follows:

- In those situations where OSHA believes there is an imminent danger to life and health (i.e., hazards that can result in death or serious physical harm), the agency will attempt an immediate inspection of the suspect facility.

- If an accident has resulted in the death of a worker or in multiple hospitalizations, employers are required to notify OSHA within twenty-four hours. This will undoubtedly result in an OSHA inspection.
- Under normal conditions (i.e., non-imminent danger or catastrophic accident situations), employee complaints that allege unsafe work practices or conditions are investigated first. Admittedly, these types of inspections take up a great deal of OSHA's time and resources. In some regions of the country, OSHA area office inspectors seldom get past this level of inspection priority in a given year. This is more a statement of limited resources and shrinking budgets than one of technical competence.
- OSHA area offices do try to carry-out routine, scheduled inspections of workplaces in their jurisdiction throughout the year. These inspections may be based upon facility size (number of employees), standard industry classification (SIC) code, type of business, and hazard level of the work performed, or any combination of these factors.
- Follow-up on repeat violations or simply revisiting a previously-inspected facility to verify abatement and compliance is the next priority. OSHA attempts to ensure, through unannounced, repeat inspections, that the actions an employer has claimed will be implemented have actually been put into place. In addition, OSHA is interested in verifying the adequacy of any abatement actions for a cited hazardous condition. In this way, OSHA not only checks on compliance, it also learns of new methods and techniques to eliminate hazards. Such information can prove invaluable at some point in the future when similar conditions appear elsewhere.

Finally, while not usually listed in the agency's official order of inspection priority, it should be noted that OSHA may also inspect a facility if a particular incident at that facility (accident, chemical spill, etc.) receives significant media attention to the point where local area OSHA officials become aware of the occurrence. In such cases, they will most likely inspect the facility as a matter of course.

To facilitate its enforcement activities and to ensure adequate access to the agency, ten regional offices have been established across the country, as shown in Figure 2-2.

OSHA also has the authority to obtain a warrant from the U.S. district courts to conduct their inspections. Chapter 8 will further discuss the inspection process and employer/employee rights during an inspection in greater detail.

Issuing Citations. OSHA citations present the details of an alleged violation, the date by which corrective action must be taken, and propose penalties. It is important to note that OSHA obtains its right to issue citations under the OSHAct itself and not by some regulation promulgated subsequent to the Act. This distinction means that, from the onset, Congress fully intended for the agency to have enforcement powers. Understanding the intent of Congress is a crucial element in OSHA compliance.

Challenging Citations. Employers also have the right to challenge or contest a citation. Specifically, within 15 days of receipt of any citation issued by OSHA, employers can file a contest to dispute the charges. If no contest is filed in the allotted time frame, then the citation will become a final order of the OSHRC. This means that the provisions of the citation, along with its allegations of violation, become a matter of fact as far as the law is concerned. If an employer unwisely decides not to respond to an OSHA *notice of alleged safety violation* or does not abate the cited hazards within the allotted time, then OSHA can issue a *failure to abate notice.* Failure to abate notices can carry additional fines and penalties over and above those contained in the original citation.

Employee's Rights. The rights of employees are also protected, under certain conditions, in the OSHAct. For instance, OSHA has the authority to prosecute employers who discriminate against any employee who exercises worker rights under the Act. These rights include the right to complain about adverse working conditions, to actually remove themselves and refuse to work under certain conditions that they believe to be unsafe, to testify against their employers on matters related to workplace safety and health, and to accompany OSHA representatives on compliance inspections. Employees can not be discharged or discriminated against for exercising any of these rights.

Figure 2-2: Geographical breakdown of OSHA regions and office locations in the United States (source: OSHA, Washington, DC).

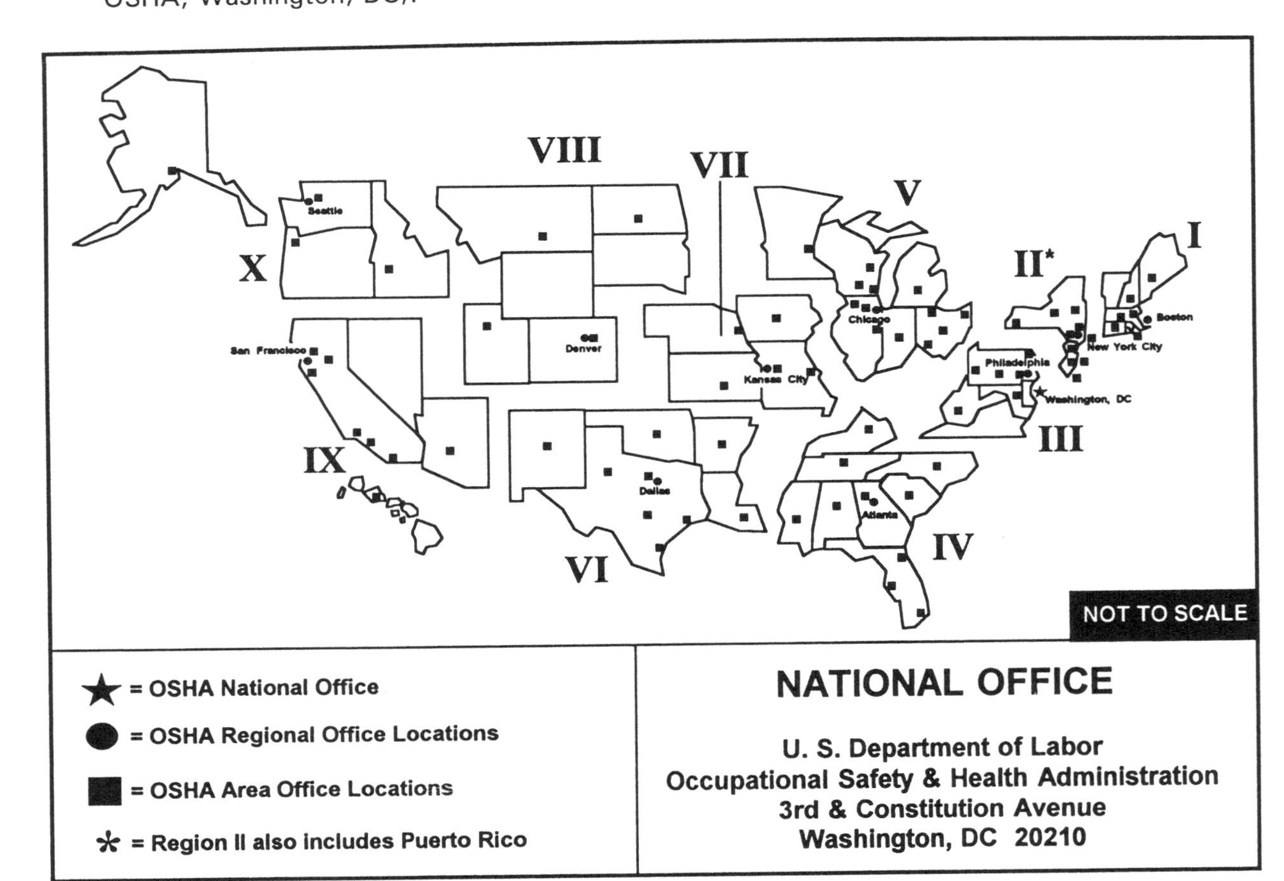

Imminent Danger. Employers should know that in cases of *imminent danger* OSHA is empowered to seek temporary restraining orders and other forms of relief from the U.S. district courts. An imminent danger situation, as mentioned earlier with regard to inspections, can cause death or serious physical harm. However, to exercise this authority, OSHA must show that the hazards created by the imminent danger situation will be eliminated through the issuance of the temporary restraining order. In reality, OSHA has not utilized this authority with any regularity. Even though organized labor representatives would prefer to see OSHA more active in responding to imminent danger situations, so much so that the ATL-CIO has lobbied Congress to grant OSHA a free hand in such situations, to this date OSHA is still not very active in pursuing temporary restraining orders in response to imminent danger in the workplace.

Criminal Prosecution for Willful Violations

When employers and their representatives cause the death of an employee through *willful* violations of regulations or standards, OSHA can recommend criminal prosecution (see Chapter 11).

OSHRC's Role: Adjudicating Contested Cases

OSHRC has the power to affirm, modify, vacate OSHA citations, or take other actions to afford relief. OSHRC hears and decides cases where citations or penalty notices (including those for failure to abate) are contested. The OSHAct specifies that OSHRC proceedings are to be governed by the APA's formal adjudication provisions. OSHRC must hold formal hearings wherein witnesses present testimony under oath and can be cross-examined.

OSHRC consists of three commissioners appointed by the President and confirmed by the Senate. The Commissioners serve six year terms. The Chairperson appoints Administrative Law Judges (ALJs) to preside over formal hearings under the Federal Rules of Civil Procedure. This guarantees a standardized approach and ensures equal administration of the law. OSHRC rules are provided in 29 CFR 2200.

As stated previously, the OSHRC and OSHA are located in separate and distinct branches within the executive branch of government. This division was intentional to ensure the fare separation or power and responsibility. While the OSHRC is responsible for impartial application of the law, employers should note that contests arising solely on matters pertaining to standards interpretation, the OSHRC (and appellate courts as well) are required to defer to OSHA's reasonable interpretation of standards.

NIOSH's Role: Researching Safety and Health Issues

NIOSH is charged with conducting research in occupational safety and health and supplying OSHA with standard recommendations. NIOSH can require employers to compile information for these research initiatives. In response to requests from employers, employees, or other representatives, NIOSH conducts health hazard evaluations (HHE's) of workplaces, a source of valuable data in addition to helping solve existing on-site difficulties. If necessary, NIOSH can issue subpoenas for employer's documents in addition to requesting warrants from the courts.

However, NIOSH has no regulatory or standard setting authority. They can only provide recommended criteria for standards. It is only when OSHA takes action on these recommendations and promulgates a standard that NIOSH recommendations become law.

What Are Employers' and Employees' Responsibilities under the OSHAct?

Chapter 6 and Chapter 7 will discuss employer compliance and employee rights and obligations, respectively, in greater detail. However, this section briefly explains these duties relevant to the current discussion of the OSHAct.

Complying with Standards

Employers and employees are must comply with OSHA standards. However, the are no provisions for federal sanctions against employees who violate their duty to comply. Section 5(a)(1) duties, referred to more commonly as the *general duty clause (GDC)* will be discussed in greater detail in Chapter 4.

The obligation to comply with standards is established in Section 5(a)(2).

Standards, divided into two categories, deal with individual safety and health problems, and require employers to comply. Understanding the difference between the two types of standards is important to successful OSHA compliance (see Chapter 3 for further information on standards). A *specification standard* details how to achieve compliance. It may establish levels of exposure or particulars on operating certain types of equipment. *Performance standards* establish what the safety or health objective is. It essentially provides details on minimally acceptable program requirements in areas such as safety training, recordkeeping, or communicating hazards to workers.

Accurate Recordkeeping

Employers must maintain records documenting their OSHA compliance activities. These include records regarding work related injuries and illnesses and records regarding toxic substance and harmful physical agents. Employees have the right to review these records and observe any workplace monitoring activities.

OSHA inspectors almost always review an employer's accident and injury records and other documents to obtain an overview of the types of accidents (and their causes) which are occurring in the workplace. OSHA recordkeeping requirements, as discussed more fully in Chapter 5, are therefore an extremely important aspect of the compliance process. OSHA also uses information obtained from employer's records to establish inspection priorities for all inspectors.

Employers' Obligations

Employers must comply with OSHA standards, recordkeeping requirements, and the GDC. This obligation can not be transferred or delegated (e.g., by contract) from one employer to another. Furthermore, employers must permit an employee or representative (e.g., a union official) to accompany an OSHA inspector during inspection, testing, or monitoring of the workplace. The employer must post in the workplace a poster that outlines all employee rights under OSHA. In addition to this, employers must adequately warn their employees of potential workplace hazards via additional, more specific posters, labels, or color codes. Whenever workplace exposure testing (e.g., air monitoring, noise monitoring) or employee exposure testing (medical tests) is required, the employer must provide for and pay the cost of any such tests.

If OSHA cites a particular workplace, then the employer must post a copy of the citation in a prominent area near the place of violation for employees to see. Finally, the employer is required to provide employees (or their designated representatives) with a copy of their personal medical and exposure records upon request.

WHAT EMPLOYERS SHOULD KNOW ABOUT CITATIONS AND VIOLATIONS

If OSHA believes an employee has been or is being exposed to any hazard governed by an OSHA standard or which the agency believes is a *recognized hazard*, regardless if a standard exists or not (i.e., the General Duty Clause, see Chapter 4), OSHA will issue a citation against the employer. OSHA can also issue a citation and propose penalties if it finds a hazard regulated by a standard even if the employer's own employees are not exposed to the alleged hazard, as is frequently the case at construction sites and factories where there are multiple employers.

When OSHA issues a citation, it does not necessarily mean that the employer has violated a standard or the general duty clause. Citations consist of allegations, employers are free to dispute (contest).

Contesting alleged violations will be discussed more fully in Chapter 10. However, it is important to know that the courts unanimously hold that employers' compliance obligations must be an *achievable duty.* In other words, the employer must either know or, with the exercise of reasonable diligence, have the ability to learn of the existence of the alleged violative condition. In addition, feasible means for controlling the hazard in question must be available and capable of eliminating or substantially reducing the hazard.

Looked at another way, this provision also means that unpreventable instances of non-compliance with OSHA standards and/or the OSHAct itself can theoretically be forgiven during the contest process. In practice, however, OSHA rarely recognizes that a situation they believe is in violation could not have been prevented. The agency usually holds the employer strictly accountable for compliance and, while it does recognize in theory that some instances could not have been prevented, it issues citations as a matter of practice. The employer is then left with the task of proving that they had no knowledge nor, with the exercise of reasonable diligence, could have had knowledge of the hazardous condition. This is a difficult task, at best.

Which Employer Is Responsible for Multi-Employer Worksites?

Employers at multi-employer worksites may be jointly and severally liable for violations at those sites. This means that each employer has the duty to fully perform their safety and health obligations and each can be held liable, either individual or as a whole, when any employee is injured at such locations. This type of situation is routine in the construction industry, and also occurring in manufacturing, processing, and other general industry operations. Multi-employer worksites pose difficult compliance problems for both employers and OSHA. As early as 1975, a court of appeals held employers violating OSHA standards responsible for the violation even if none of its own employees were exposed to the hazard. The employer who created the hazard is held to be in the best position to abate it. But the employer could lose the ability to control the hazard when a subsequent employer is in control.

But what if the employer responsible for creating the hazard is the only employer on site with the needed expertise to eliminate the hazard (e.g., the electrician may be the only contractor able to properly abate electrical wiring hazards.) While OSHA and OSHRC take the position that a general contractor, construction manager, or owner on a multi-employer worksite has the authority or ability to correct hazards at that site, OSHA does not always cite the general contractor. Usually, OSHA will look for and cite the employer having responsibility for the hazard. Employers whose employees are exposed to hazards which the employer did not create will still be held liable unless that employer attempted to take control of and eliminate the hazard.

The non-hazard-creating employer is typically a subcontractor at the worksite. OSHRC requires that subcontractors attempt to get the general contractor to abate the hazardous condition(s) or ensure that its employees avoid exposure. In practice, this means that those employers who did not create the hazard can protect themselves from citation by either preventing their own employees from exposure or by making visible (e.g., documented) attempts to get the general contractor to eliminate the hazardous condition.

One final note on the subject of multi-employer worksites. Since the OSHAct does not impose strict liability, the non-hazard-creating employer or contractor may also defend itself on the basis that it did not know nor could not, with the exercise of reasonable diligence, have known of the presence of the hazard. Although this defense is seldom successful, employers should be aware of its potential.

EMPLOYEES' DUTIES

While the OSHAct does require employees to comply with OSHA standards, there are no provisions for enforcement against individual employees or labor unions. This is not meant to imply that employers cannot discipline employees who violate OSHA standards. In fact, employer discipline for failure or refusal to comply with OSHA standards is frequently upheld by arbitration under most labor contracts.

The Obligation to Work Safety

Also, referencing the discrimination provisions of the Act discussed earlier in this chapter, disciplinary measures taken by employers solely in response to employee refusals to comply with appropriate safety rules and regulations is not considered discrimination and is not in violation of the Act. This is an important but subtle differentiation. Employers must have the freedom to manage their enterprises. This freedom must include the right to discipline employees when conditions warrant. The Act does not attempt to undermine this right. Therefore, employees who are disciplined for refusing to comply with their employer's safety and health rules will not be granted relief under the discriminatory provisions of the Act. In fact, employers can and have successfully defended themselves against OSHA citations when they discipline employees who, during inspections, are found not to be complying with workplace rules regarding the hazard which have been communicated by the employer.

Chapter 3

OSHA STANDARDS, THE RULEMAKING PROCESS, AND ENFORCEMENT

OVERVIEW

The OSHAct establishes the objective of assuring workplace safety and health. However, it does not address which workplace hazards are to be regulated or the methods to be used to reduce or eliminate them. But it does provide OSHA with the authority to adopt and promulgate standards to address occupational safety and health issues. During the first two years of its existence, OSHA was authorized to review and adopt existing national consensus standards without going through any formal rulemaking process. These standards have come to be known as *Section 6(a) Standards* (after the section of the Act that allowed such action) or, more commonly, as *1971 Base Standards.* Any new hazards discovered after the promulgation of the 1971 Base Standards are addressed and subsequent standards are developed under the *rulemaking* provisions of Section 6(b). There are also provisions which allow OSHA to promulgate *emergency temporary standards* by foregoing Section 6(b) requirements. However, such standards only carry the force of law for a six month period and are usually invalidated by the courts at some time after that.

The *rulemaking process* follows the baseline procedures established under the Administrative Procedures Act. Specifically, there must be a notice of proposed rulemaking (NPR) published in the Federal Register. OSHA must allow for a hearing during which any and all parties interested in the action can participate. When the rule is finally issued, there must also be published a *preamble* to the standard which provides great detail on the development of the final rule and explains the intent of OSHA in promulgating the rule. Post-hearing procedures allow for challenges of the rule to be lodged by employers and other interested parties based upon the rule's validity, feasibility, and practicality.

There are various types of standards which OSHA may pursue. While specification standards address specific hazards or concerns, performance standards are goal oriented and broad in scope. Each of these were briefly discussed in the previous chapter. In addition, *horizontal standards* cover all of general industry, while *vertical standards* focus on specific industries, such as construction.

Under certain conditions, OSHA may issue *variances* to employers which will allow them to avoid compliance with a given standard for a specific period of time (temporary variance), or indefinitely (permanent variance). Rarely, a variance may also be issued against any standards to preserve the *national defense* (such variances are seldom more than six months in length). Regardless of the type of variance, the employer or petitioner must show OSHA that they can still provide a level of safety and/or health that is consistent with the intent of the standard for which the variance is being sought during the life of the variance.

In recent years, OSHA rulemaking has focused increasingly on employee education and training and employer self-inspections and audits. If this is any indication of a trend, then future actions by the agency might be expected to contain similar provisions.

INTRODUCTION

Intended by Congress to provide workers with safe and healthful working conditions, the primary objective of the OSHAct did not specify which hazards are to be regulated or how they were to be controlled, instead granting OSHA the authority to adopt occupational safety and health standards. These standards then become the instruments by which compliance with the Act's primary objective can be achieved. As new hazards are discovered and identified, new standards can be promulgated to address exposure reduction or elimination.

The term *occupational safety and health standard* is defined in the OSHAct as a standard which requires conditions, or the adoption or use of one or more practices, means, methods, operations, or processes, reasonably necessary or appropriate to provide safe or healthful employment or places of employment. In other words, standards must do whatever it takes, within reason, to assure a safe and healthy workplace.

Employers must therefore comply with occupational safety and health standards.

Although the employer's duty to comply with standards is specified in Section 5(a)(2) of the Act (the general duty clause, see Chapter 4), this duty to comply is not entirely absolute. Rather, the Act imposes an *achievable duty* of compliance on employers. This means that employers are not required to *guarantee* the safety of their workers under *all* possible working conditions and at *all* times. It simply means that employers must take reasonable and appropriate measures to ensure the safety of their workers so far as possible. Standards are developed and promulgated to assist employers in this effort. Standards establish the *minimum-acceptable* levels of safety under specific workplace conditions.

THE RULEMAKING PROCESS

Standards and How They Are Established

Figure 3-1 shows the sources from which safety and health standards are derived. As stated previously in this text, OSHA derives authority to promulgate standards under Section 6 of the OSHAct. Initially (i.e., in 1971), OSHA was authorized to adopt national consensus standards, such as those already developed by organizations such as the American National Standards Institute (ANSI), and established federal standards (like those in the Walsh-Healey Act) within the first two years after the Act became law (Section 6(a) of the Act). These standards were to establish a federal regulatory foundation upon which to build new requirements and standards as situations warranted.

Congress also knew that these consensus standards would not be enough to meet the requirements of the Act into the future. New work situations, technologies, methods, and improvements would most likely require additional standards to address any new hazards that arise. Therefore, Section 6(b) provides OSHA with the authority to promulgate new OSHA standards, to update, as required, the national consensus standards, and to establish new federal standards. In addition, OSHA is authorized to promulgate emergency temporary standards without the use of normal rulemaking procedures. Such standards are only effective for a maximum of six months. Section 6(d) provides the authority for OSHA to issue permanent variance from standards. Each of these actions are discussed separately below.

Figure 3-1: The various sources from which occupational safety and health standards are derived.

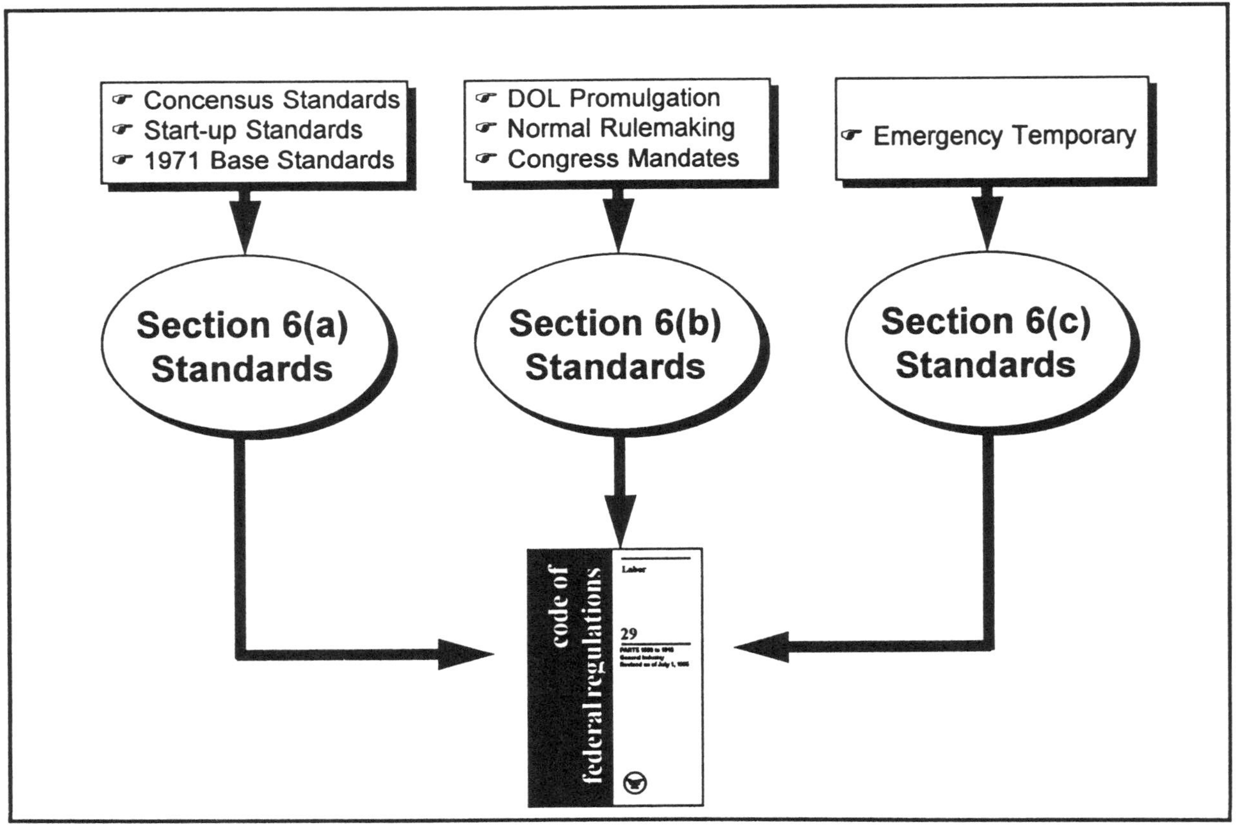

- Concensus Standards
- Start-up Standards
- 1971 Base Standards

Section 6(a) Standards

- DOL Promulgation
- Normal Rulemaking
- Congress Mandates

Section 6(b) Standards

- Emergency Temporary

Section 6(c) Standards

code of federal regulations

Labor

29

Section 6(a) Standards. Industry and federal occupational safety and health standards in existence in 1971 were adopted under the authority of Section 6(a) of the OSHAct. These are sometimes referred to as 1971 Base Standards or Section 6(a) Standards. For a period of two years after the Act became effective, these standards were adopted on an almost continuous basis without regard for the normal rulemaking processes outlined in Section 6(b).

Consensus Standards. The OSHAct provided a rather convoluted and extensive definition of the term *consensus standard* (see 29 U.S.C. Section 652 (9) for the full definition). For the purpose of this text, the definition is simplified. In general, a consensus standard is any occupational safety and health standard that:

- Has been adopted by a nationally recognized, standards producing organization;
- Is determined to be of specific interest to those affected by the scope and provisions of the standard;
- Has been adopted and accepted by those affected;
- Was developed in a manner which afforded those interested the opportunity to participate in its development;
- Has been designated as a consensus standard by OSHA after much consultation with other Federal Agencies.

Using this definition, OSHA determined that a number of ANSI and National Fire Protection Association (NFPA) standards, for example, would qualify as Section 6(a) Standards. It should be noted that OSHA was not authorized to make any changes or amendments to these standards or to any established federal standards at the time of their adoption. Any such changes would require OSHA proceedings under Section 6(b).

Established Federal Standards. The term *established federal standards* is defined as any occupational safety and health standard established by any federal agency in effect or contained in any Act of Congress in force on 29 December 1970. This provision allowed the adoption of established federal standards in the construction, shipbuilding, and ship breaking industries.

Employers should know that, because there was no opportunity for public comment on Section 6(a) Standards, those cited for allegedly violating these standards are permitted to challenge their validity during enforcement proceedings.

Section 6(b) Standards. Under normal conditions, OSHA issues, modifies, and/or revokes occupational safety and health standards under the rulemaking procedures of Section 6(b) of the Act. Intended to create a way to modify the Section 6(a) "start-up" standards and generate new worker protection regulations. The Section 6(b) rulemaking authority is somewhat similar to the informal notice and comment rulemaking provisions of the Administrative Procedures Act (APA), as discussed in Chapter 1. However, OSHA has determined that additional requirements are necessary when promulgating standards under Section 6(b).

PROMULGATION OF A STANDARD

Risk Analysis, Feasibility Studies, and Cost-Benefit Analysis

Specifically, the procedural requirements for promulgating, amending, or revoking an OSHA standard require that the process begin with notice of proposed rulemaking (NPR), and a hearing, if requested. When OSHA promulgates a standard it must be supported by substantial evidence developed on record during the rulemaking process. For any new standard, OSHA must perform a *risk analysis* to determine if a significant risk of health impairment exists under current regulations before issuing the standard.

For *health standards*, OSHA must perform a *feasibility study* to determine if the proposed standard is practical for the exposure under consideration as well as from an implementation perspective. For *safety standards*, there must be a *cost-benefit analysis* to determine that the costs of implementing the standard bear a reasonable relationship to its subsequent (anticipated or expected) benefits.

What does all this mean to the employer? For whatever it is worth, employers should know that by the time a new standard appears in the FR or CFR, OSHA has already conducted numerous studies to ensure that the need for the standard outweighs any potential costs associated with its implementation. This is, of course, little consolation to an employer who may have to spend literally hundreds of thousands of dollars and perhaps more to achieve compliance with a given new standard.

Section 6(c) Emergency Temporary Standards

Emergency temporary standards (ETS), are issued under Section 6(c) without using the notice and comment rulemaking procedures. OSHA can issue an ETS any time it determines that an ETS is necessary to protect employees who are being subjected to extreme danger from exposure to known toxic or physically harmful substances or agents. Specifically, in order to promulgate an ETS, there must be a "danger of incurable, permanent, or fatal consequences to workers exposed, as opposed to easily curable and fleeting effects on their health." An ETS is effective for six months following publication in the Federal Register.

It is important to note that use of an ETS does not mean OSHA can avoid the public comment and hearings of Section 6(b) rulemaking procedures. What the Act does allow is for OSHA to immediately commence normal rulemaking procedures in conjunction with the release of the ETS, using the ETS as the proposed rule. The Section 6(b) rulemaking should be completed within six months, making the ETS (possibly in revised form) a final standard through the normal rulemaking process. In practice, however, OSHA seldom moves that quickly through the process and the ETS is typically invalidated by the courts at some point after the six month expiration date.

How the Rulemaking Process Works

OSHA may initiate the Section 6(b) rulemaking process after reviewing petitions received from interested parties, recommendations from other federal agencies, OSHA advisory committees, or on its own initiative. The National Institute for Occupational Safety and Health (NIOSH) provides OSHA with health (primarily) and safety standard recommendations.

The Comment Period. OSHA publishes an NPR in the Federal Register and invites public comment when it determines that a standard should be issued, revised, or revoked (see Chapter 1, Figure 1-2). If the agency subsequently decides that a standard is not needed, it must announce in the Federal Register its reasons for this determination. While OSHA must allow a minimum of thirty (30) days for the public to respond with comments once the NPR has appeared in the Federal Register, the period for filing initial or opening comments is usually set at sixty (60) days.

During this period, interested parties can offer opinions, objections, and request a hearing to further discuss the proposed standards. A hearing must be held if any objections are filed.

All initial comments received by OSHA are available for review by any interested party. OSHA often hires outside contractors to prepare unbiased opinions on the feasibility and potential costs of the proposed standard after evaluating the initial comments filed. The contractor also testifies at any public hearing regarding the proposed standard.

The Hearing. The hearing process is typically informal. Witnesses are not sworn in and cross-examination may not be allowed. An Administrative Law Judge (ALJ) presides over the hearing, and a transcript of the proceedings is prepared.

At the conclusion of the hearing, the ALJ certifies the record, exhibits, written comments, and transcripts to OSHA, for a final determination. Hearing officers will often allow post hearing comments and briefs to be filed before closing the record.

Promulgating the Standard. OSHA is then supposed to publish the final standard in the Federal Register within sixty days after the close of the notice and comment period or hearing. In actuality, OSHA usually spends several months preparing the final standard and its preamble. The preamble is an integral and important part of any OSHA standard. It is required by law and is used to explain the decisions made by OSHA and discuss the evidence relied upon to support the final standard. OSHA's initial interpretations concerning the intended meanings of the various provisions of the final standard are also provided in the preamble. Hence, while nothing contained in the preamble carries the force of law, the information provided there does offer a great deal of valuable data concerning the standard itself.

Making sense of OSHA standards is seldom easy. In terms of compliance, employers should always look to the preamble of any new standard to ensure understanding of OSHA's intentions.

Requesting a Stay. Anyone who may be adversely affected by a new, modified, or revised standard can file a petition for review with the U. S. Court of Appeals. The petition must be filed within sixty (60) days after the promulgation, revision, or modification of the standard in question. The calendar begins from the date of publication in the Federal Register. The party may seek a *stay* of the standard (i.e., a delay in its enactment), pending the court's review. If granted, the stay will render the new standard, or portions of it, unenforceable during the review. To be successful when filing a petition for a stay, the party must establish that:

- There is a substantial likelihood that the petitioner will prevail on the merits of the stay;
- The petitioner will be irreparably harmed if the stay is denied;
- Other potential parties to the proceeding will not be substantially harmed by the stay, and;
- The stay will not interfere with the public interest.

ENFORCEMENT

Types of Standards and Their Enforcement

Generally, OSHA standards are designed to ensure workers are provided with a safe and healthy workplace. There are basically two types of standards OSHA can enact to accomplish this objective. *Safety standards* are designed to protect employees from hazards such as slips, trips and falls, lacerations and amputation from using machinery, fire hazards, and so on. *Health standards* generally establish measurements for worker exposure to hazardous agents. Such hazards frequently involve the potential for long-term adverse health effects (exposure to lead, noise, asbestos, silica, radiation, vibration, etc.).

OSHA standards can differ somewhat, depending upon their intent and purpose. As shown in Figure 3-2, the types of standards can be divided into two general categories. There are *specification and performance standards* and *horizontal and vertical standards.*

Figure 3-2: The types of standards.

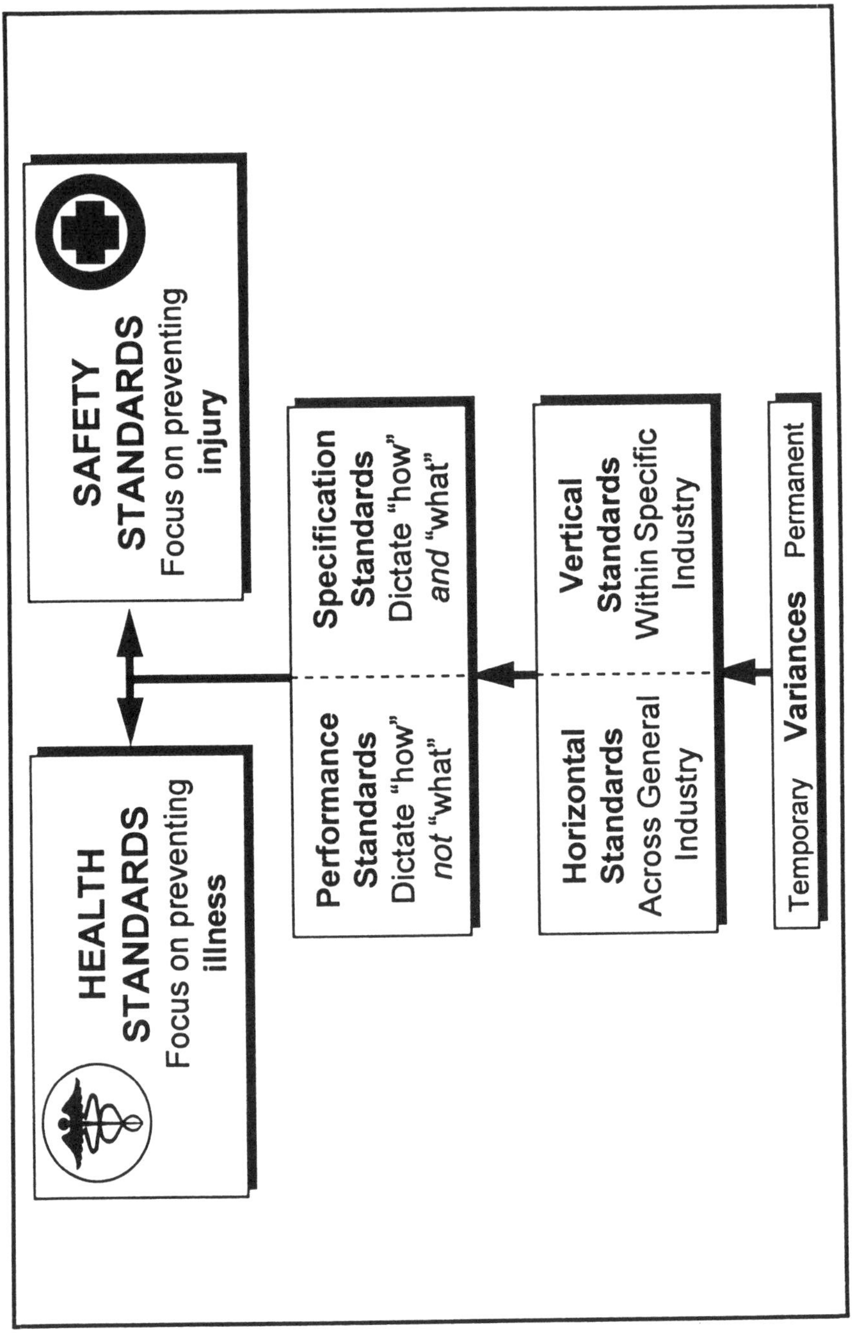

Specification Standards. First introduced in Chapter 2, a specification standard establishes the specific methods employers must utilize for hazard abatements. For example, OSHA's specifications for guardrail protection in general industry require that open-sided floors, wall openings, and floor openings be guarded by a rail system meeting the specifications established in the standards (See 29 CFR 1910.23). Whenever OSHA issues a citation for an alleged violation of a specification standard, OSHA has the burden of proof to show only that:

- The standard applies to the cited employer;
- The standard's requirements were not met;
- Employees were exposed to hazards as a result; and
- The employer knew or, with the practice of reasonable diligence, should have known of the noncompliance conditions' existence.

The Burden of Proof. The *burden of proof* is very clear. OSHA establishes a *prima facie* (i.e., before further examination) case supporting the citation when these four points are established. The employer must then establish a case by way of avoidance. The employer may sometimes claim that "employee negligence or misconduct," resulted in the noncompliance condition. In using this method of defense, an employer claims that all the requirements under the specification standard were fulfilled. But, because an individual employee (or group of employees) acted against the employer's orders, employees were exposed to the hazard which resulted in the citation.

In the real world, however, this approach is not always successful. The fact that employees decide not to observe OSHA standards is not an acceptable defense. It is always the employer's responsibility to discipline employees and to take steps to ensure that they comply with established safety and health procedures. Also, employers may not delegate the responsibility of safety compliance in the workplace to their employees. Other defenses, depending upon the circumstances of the citation, include that compliance was simply not possible, practical, or feasible under the specific conditions (including time constraints) in the workplace at the time of the citation. The employer may also claim that the cited standard was invalid as applied to the facts of the circumstance. Whatever the approach, it is important to note that the burden of proof in all such cases will be on the employer to demonstrate to the court that OSHA was wrong in alleging the cited conditions.

Performance Standards. Recalling the discussion from Chapter 2 regarding employer and employee duties to comply with standards, a "performance" standard is goal oriented. While it does not stipulate how such objectives are to be achieved, it does establish what the particular goals will be.

For example, the personal protective equipment standard requires that employees be protected by the use of this equipment whenever necessary because of hazards that are encountered on the job. But it does not specify how such a goal will be achieved. Even though it contains no specific requirements, OSHA has used this standard to require employees to use rubber insulated gloves, respirators, goggles, face shields, and so on, when working with electrical equipment.

Goal-oriented performance standards are typically vague or broad in nature. This provides OSHA the ability to evaluate an employer's attempts to achieve the goal without regard to any specific or established requirements. Conversely, it should be noted that performance standards provide employees greater flexibility to create a compliance program to meet the goals of the standards. Generally speaking, OSHA will attempt to adopt a performance standard over a specification standard where possible.

Horizontal Standards. Regulating occupational safety and health issues across industry lines, "horizontal" standards are usually applicable to general industry (as opposed to the construction, maritime, or other specific industry) and apply to all covered employees. An excellent example of a horizontal standard is the Hazard Communication Standard (29 CFR 1910.1200), which applies to all employers covered by the OSHAct. Standards provided in 29 CFR 1910 ("General Industry") are typically established as horizontal standards.

Vertical Standards. "Vertical" standards apply to specific industries. Standards established in 29 CFR 1926 cover the construction industry. Such standards cover many of the same hazards found in general industry horizontal standards, which may lead to confusion in compliance efforts. There are also many industry-specific hazards which must be addressed solely by the vertical standard.

In an attempt to resolve any confusion, OSHA states that when a particular standard specifically applies to a condition, practice, means, method, operation, or process, it prevails over any general standard which might otherwise apply. In other words, a vertical standard will take precedence over a similar horizontal standard if a conflict between the two should arise and the vertical standard is directly applicable.

But it also means that employers must still abide by the horizontal standards in cases when those standards are not preempted by any similar vertical standard. In simpler terms, it is safe to assume that a given condition in a specific industry will most likely be covered by either a vertical standard or a horizontal standard and the employer must still take action to ensure a safe workplace. For instance, if a topic is covered by the general industry electrical safety standard (29 CFR 1910, Subpart S), then the general industry standard will apply to the construction job even if it is not covered in the electrical safety standards work (29 CFR 1926, Subpart K).

What Are Variances?

Employers can petition OSHA for an order granting a variance from any standard. Currently, there are two main types of variances: *temporary* and *permanent*.

Temporary Variances. Temporary variances allow employers to operate for limited periods of time under certain noncompliance conditions. OSHA will not permit any temporary variance to remain in effect for a period longer than needed by the employer to achieve compliance, or for one year, whichever is shorter. The employer may petition for renewal of the variance, but not more than twice.

An employer's temporary variance application can be filed at any time after the standard's promulgation. A temporary variance from a new OSHA standard may be obtained if the employer:

- Is unable to comply with the standard prior to its effective date because workers, facilities, or equipment are not available;

- Is using all possible measures to protect employees from the hazards covered by the standard; and
- Has in place plans for achieving compliance as quickly as possible.

Permanent Variances. Under certain circumstances, an employer may obtain a permanent variance from an OSHA standard. As the name implies, these variances are granted for an indefinite time and are only issued if and when OSHA finds that the workplace is as safe and healthful as it would be under the standard in question. For permanent variance applications, a formal hearing may be held at which the petitioning employer has the burden of proving the merits of the request. OSHA seldom grants a permanent variance from a performance standard, since such standards normally allow the employer to utilize whatever means necessary to attain the performance requirements of the standard.

If a permanent variance is granted, it is usually and purposely narrow in scope. It will probably specify the alternative practices or conditions the employer must meet. It is also important to understand that a permanent variance can be modified or revoked by OSHA at any time after its issuance. Such action can be done by OSHA on its own motion or following a proceeding initiated by an affected employee. No application for revocation or modification of the permanent variance can be made, however, unless the variance has been in effect for a least six months.

National Defense Variances. On rare occasion, a variance may also be obtained from OSHA for reasons of safeguarding or preserving the *national defense*. OSHA has the discretion to "allow reasonable variations, tolerances, and exemptions to and from any or all provisions of the Act" to "avoid serious impairment of the national defense" (29 USC, Section 665). If a national defense variance is in effect for longer six months, affected employees must be advised and granted a hearing if desired.

Challenging OSHA Standards

Under Section 6(f) of the OSHAct, employers can challenge new OSHA standards within sixty days of promulgation. If they are supported by "substantial evidence," as provided by OSHA, the standards will be affirmed. These challenges are only one option available to employers for contesting a standard's validity. Standards can also be challenged during enforcement proceedings. For example, a cited employer may claim that the standard upon which a citation is based is *invalid* because compliance is not technologically or economically feasible, or because there were procedural deficiencies in the standard's issuance. Challenging an OSHA standard is seldom easy, but it can be accomplished to the employer's benefit. To be successful in such cases, employers must understand the particulars of the case against them and bring forward information or evidence to prove that OSHA was in error when it alleged the citation.

The Future of OSHA Rulemaking

It is not possible to elaborate on the future of the agency itself, especially considering that political agendas often dictate the role a federal agency will play from one presidential administration to the next. However, based upon recent rulemaking actions by OSHA, it is possible to hypothesize on the direction or focus future standards may take in the coming years.

A review of relatively recent standards such as the Lockout/Tagout Standard (29 CFR 1910.147), the Bloodborne Pathogens Standard (29 CFR 1910.1030), and the Personal Protective Equipment Standard (29 CFR 1910.132) indicates that OSHA is focusing more heavily on employee education and training than ever before. This may have begun most visibly with the promulgation of the Hazard Communication Standard (29 CFR 1910.1200) in the early 1980s. Since that time, new and proposed standards require employers to ensure that their employees are properly and adequately trained for the hazards they may encounter in the workplace. The proposed Ergonomic Standard is a perfect example. If it is ever made final, it is certain to contain very precise requirements aimed at employee education and training.

There are other possible areas on which OSHA may place increasing importance during future rule-making actions. These may include an increased focus on employer self-inspection and audits for safety and health compliance. OSHA's current Voluntary Protection Program (VPP) is an example of the importance the agency places on employer self-monitoring and inspection.

Congressional funding and budgetary adjustments will force the agency to operate more efficiently. Much like corporate America, OSHA will undoubtedly be required to "do more with less" in future years of diminishing budgets and Congressional scrutiny. For example, in fiscal year 1996, the Congress had proposed a $54 million decrease in OSHA's annual operating budget of roughly $312 million (which is nearly 1/30th the budget of the EPA). This translates to a decrease in OSHA personnel of approximately 650. OSHA's final 1996 budget was nearly $305 million. A $340 million FY1997 budget was initially proposed by the Clinton Administration, with Congress proposing a merger between OSHA and MSHA, resulting in a $90 million in savings. The final 1997 OSHA budget was just under $326 million.

As a result of such changes, OSHA has been in the process of changing itself. According to the Assistant Deputy Secretary of Labor for OSHA, the agency will change from a *process oriented* organization to a *results oriented* organization. Instead of being measured simply by the number of inspections it has performed in a given year, it will measure its success on the results of those inspections. Specifically, the agency will look at the improvements made in workplace safety and health as a result of OSHA's involvement. This means that employers will have a greater opportunity to work with the agency as partners in occupational safety and health rather than the often adversarial relationships that has been common in the past. Such changes remain to be seen.

Chapter 4

THE GENERAL DUTY CLAUSE

OVERVIEW

When Congress enacted the OSHAct in 1970, it realized that the new Occupational Safety and Health Administration could never promulgate safety and health standards to cover *all* possible or potential hazards under *all* conditions that may occur in *all* types of work environments. To fill this gap in the standards, the OSHAct contains *a general duty clause (GDC)* which requires employers to provide a safe and healthy workplace that is "free from recognized hazards that are causing or are likely to cause death or serious physical harm" to employees. In the absence of a specific standard that addresses a specific hazard, the GDC imposes general requirements upon the employer.

There are certain *key elements* of the GDC that must exist in order for a citation under this clause to be considered plausible. First and foremost, the hazard being cited must be a *recognized hazard*. OSHA can not expect an employer to protect against hazards that are not recognized as such. Recognition can be substantiated in a variety of ways, and OSHA must prove to the courts that the hazard was, indeed, recognizable at the time it was cited. For example, *personal recognition* by the cited employer can be verified through review of employer statements about the hazard, company documents and rules, or previous instances where the employer is known by the employees to have addressed the hazard.

Another factor under the recognition element of the GDC is *industry recognition* of the hazard's existence. Under this approach, if OSHA can show that the employer's industry was aware of the hazard's existence, then even if the cited employer did not know about it, they can still be cited for a GDC violation because they should have known.

Foreseeability is another element of the GDC that can prove recognition. If OSHA can show that the cited employer had reason to believe the conditions which caused an accident were hazardous, then the employer should have foreseen the likelihood that an accident could occur and recognition is therefore established.

Once it is established that a hazard is recognized, OSHA must also show that such hazards are causing or are likely to cause death or serious physical harm. If this has not been established, then citation under the GDC is not appropriate. It is important to note that the occurrence of an accident is not essential to proving that the likelihood of death or serious physical harm is possible. OSHA must only show that, if an accident were to occur, the results would be death or serious physical harm.

Another aspect of the GDC is the *feasibility* factor. In any citation of the GDC, OSHA must show that feasible means of abating the hazard exist. Abatement strategies must be both technologically and economically feasible to implement.

Finally, it must be clearly understood that citation under the GDC cannot be used where a specific OSHA standard already exists to address the same hazards or conditions. OSHA cannot use the GDC to strengthen a citation against some existing standard, nor can OSHA utilize the GDC as a method of enforcing non-mandatory recommendations (i.e., those identified with a "should" rather than a "shall" or "must").

While an employer is always expected to provide whatever measures necessary to ensure a safe and healthy workplace, they should never assume that simple compliance with the minimum requirements of a given standard satisfy the intent of the OSHAct. Many times, prudent management requires exceeding a standard's requirements to ensure employee safety and health. This is a general duty of all employers and citation under the GDC is very likely if the employer fails to meet this duty, regardless of their level of compliance with established standards.

INTRODUCTION

Thus far, this text has established that an employer's principal duty under the OSHAct is to comply with occupational safety and health regulations. This might suggest that a regulation must first exist before an employer is expected to comply with it. Using this logic, if a hazard exists but no regulation or standard has yet been issued to address it, then an employer might not be responsible. Already this logic sounds flawed, and it is. Congress acknowledged from the onset that all safety and health hazards would not or could not be regulated by specific standards. OSHA would not be able to issue a standard for each and every possible and potential condition in every workplace under all working conditions. To address this issue, Congress included in the OSHAct a provision covering hazards, which although 'recognized,' were not covered by standards.

Located at Section 5(a)(1) of the Act, the *general duty clause (GDC)* provides that:

> Each employer shall furnish to each of his employees employment and a place of employment which are free from recognized hazards that are causing or are likely to cause death or serious physical harm to his employees.

Section 5(a)(2) requires employers to comply with occupational safety and health standards promulgated under the Act. In simple terms, this clause contains both the employer's duty to comply with OSHA standards and, in the absence of any specific standards, the general duty to provide a safe and healthful working environment.

It should also be noted that employees have a general duty under the Act as well. Under Section 5(b), it is stipulated that:

> Each employee shall comply with the occupational safety and health standards and all rules, regulations, and orders issued pursuant to this Act which are applicable to his own actions and conduct.

In actual practice, however, it is more common to find that general duty clause violations focus on an employer's failure to provide a safe or healthy work environment rather than an employee's failure to comply with the general duties of Section 5(b). While the Act specifically states that each employee must comply with occupational safety and health requirements, no mechanism to force compliance by employees exists under the Act. Therefore, except in cases of unauthorized or disobedient employee action, ultimate responsibility for compliance still lies with the employer (regardless of the language of Section 5(b) of the GDC).

While OSHA does not intend to make the employer the absolute insurer of the conduct of all employees and virtually liable for all acts of omission or commission, the agency does take the position that because the employer controls the work environment, the responsibility to the employee under the GDC is very high.

EVOLUTION OF THE GENERAL DUTY CLAUSE

Since the OSHAct went into effect, OSHA has used the GDC to differing degrees, depending on the presidential administration. For instance, in 1973 and 1974, under President Nixon, OSHA issued GDC citations less than two hundred times per year. This number rose to 720 per year during the Ford years. GDC citations reached their peak during the Carter Administration when, in 1978 alone, they totaled over 3,800. At that time the GDC was OSHA's most frequently cited provision. It was even suggested by some in Congress, as well as those in industry, that OSHA was abusing its use of the clause. Subsequently, in 1981, the first year of the Reagan Administration, the use of GDC citations decreased 52 percent. The use of the clause to cite employers has somewhat stabilized since the peak in 1978. However, they continue to be issued to this day and employers must be aware of the GDC's implications. While application of the GDC has fluctuated over the years, its importance cannot be overstated.

In early OSHRC decisions regarding the GDC, it was established that employer liability under the GDC provision does not require the occurrence of an accident or employee injury. Rather, liability is a condition based only on whether the employer has exercised reasonable diligence in providing employees with a place of employment free from recognized hazards. Hence, the question of employer liability under the GDC falls upon the definition of the concept of *recognized hazards* in any particular case (Figure 4-1). Recognized hazards include an employer's knowledge as well as the generally recognized knowledge of the industry in which the hazard(s) exist(s). For further clarification, refer to OSHA's Field Operations Manual (FOM). The FOM states that a hazard is "recognized" if it is a condition that is:

- Of common knowledge or generally recognized in the particular industry in which it occurs; and
- Detectable by means of the senses (sight, smell, touch, and hearing) or, is of such wide recognition as a hazard in general industry that even if it is not detectable by the senses, there are generally known and accepted tests for its existence which should make its presence known to the employer.

Figure 4-1: Determining if a General Duty Clause violation exists.

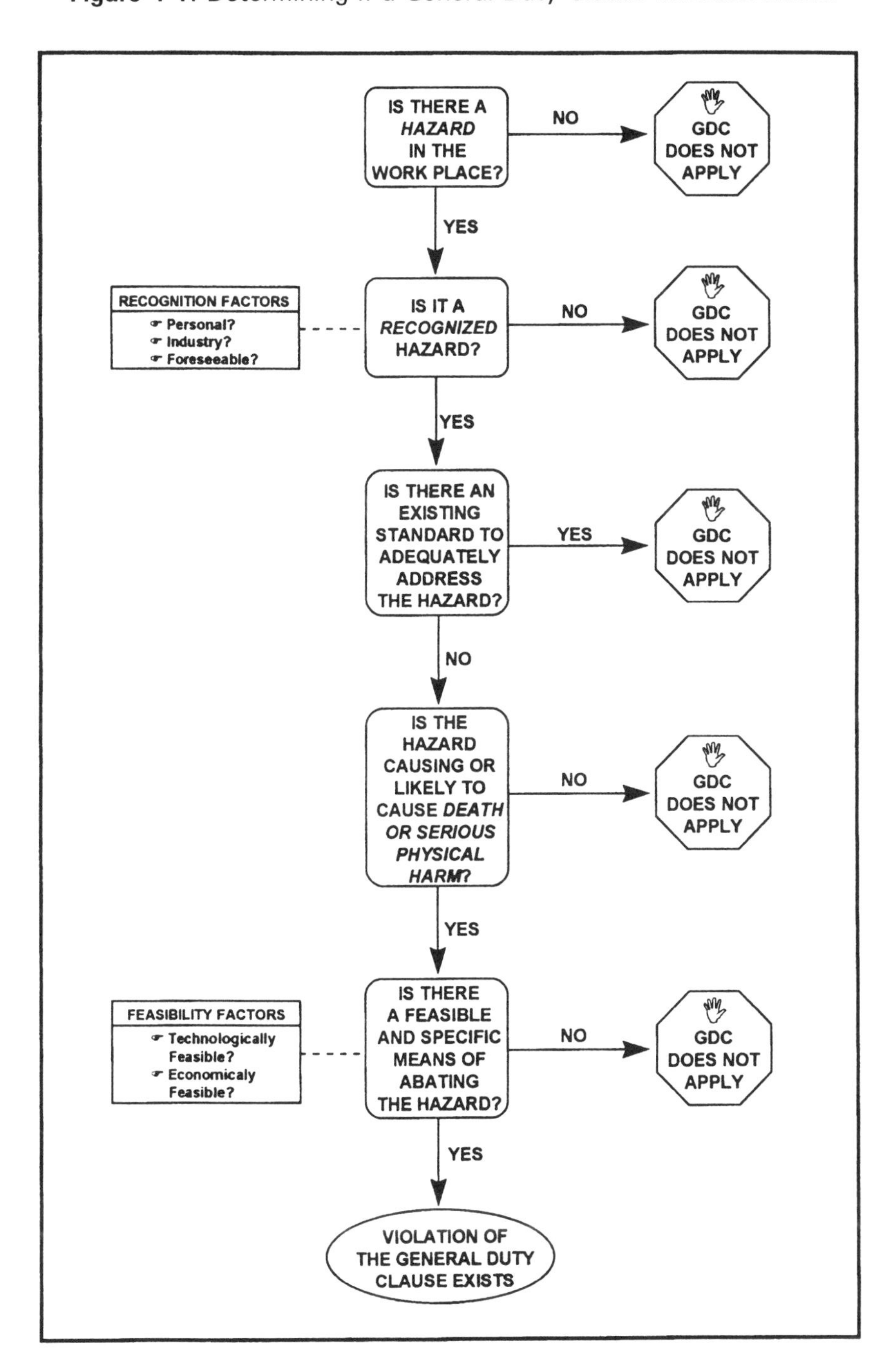

The OSHRC has favored the above definition on numerous occasions. The Commission has found that serious hazards, not always obvious to the senses, can be detected by instrumentation and tests and that this is within the scope of the GDC. Furthermore, if hazards not discernible to the senses are suspected, detection instruments should be used. So, employers are not only expected to comply with the GDC when hazards are known to exist, they are also required to comply if they should have known of the hazards' existence.

COMPLYING WITH THE GENERAL DUTY CLAUSE

It should be understood that the GDC does not impose absolute liability on the employer whenever an employee, for whatever reason, is injured or killed on the job. Congress intended the general duty of employers to be an achievable one and the hazards that can give rise to a general duty violation must not only be *recognizable,* they must also be *preventable.*

In order to substantiate a violation of the GDC, OSHA must prove a specific set of conditions or circumstances. The agency must show that:

- The employer failed to render the workplace free of a hazard;
- The hazard was *recognized* either by the cited employer or generally known as such within the employer's industry;
- The hazard was causing or likely to cause death or serious physical harm; and
- There was a feasible means by which the employer could have eliminated or materially reduced the hazard.

Also, for a charge made under the GDC to hold firm, OSHA must specify the particular steps an employer should have taken to avoid citation and demonstrate both the feasibility and the likely effectiveness of the steps.

An analysis of these requirements reveals the following key intentions of the GDC:

- The duty to protect the employee is imposed upon the employer by the GDC;
- The hazards against which the employer must protect the

employee necessarily include those of which the employer has specific knowledge;

- If an employer knows that a particular safety standard is inadequate to protect against a specific hazard, or that the conditions in the employer's workplace are such that the safety standard will not adequately deal with the hazard, then the employer has a duty to take whatever measures necessary (over and above the standard) to safeguard the employee.

In other words, if an employer knows that a specific standard will not protect against a particular hazard, then the duty to protect employees under the GDC will not be alleviated no matter how faithfully the standard is observed. Clearly, *knowledge* of a standard's inadequacy is the key to this aspect of GDC applicability. If the same employer absolutely relied upon compliance with an existing standard to protect employees and *had no knowledge* that the standard was inadequate, then the employer may be absolved from the liabilities associated with and arising from an injury, illness, or accident due to a hazardous exposure. Compliance in the latter case does not relieve an employer of the protection duty under the GDC; rather, it satisfies that duty. The key here is that the employer truly believed that compliance with the standard was adequate to protect against a hazard.

HAZARD RECOGNITION

It has been established that employer recognition of a hazard is key to a GDC citation. In citing the GDC, OSHA must not only state *what* the hazard is, they must also prove that it is *recognized* as likely to cause death or bodily injury. Recognized hazards which do not involve serious injury or the possibility of death are not covered by the GDC. Hence, the question to be asked is whether the recognized hazard will produce serious physical harm or death should an accident resulting from exposure to that hazard occur. According to the OSHRC, a *safety hazard* is a condition that creates or contributes to an increased risk that an event causing death or serious bodily harm will occur.

What constitutes a *recognized hazard* under the GDC has been the subject of much litigation. As stated previously, the courts have decided the term includes an employer's actual knowledge (i.e., *personal recognition*) of the hazard, as well as common knowledge of the condition within the employer's industry (i.e., *industry recognition*). Hence, even if the cited employer did not know of the condition, he/she should have, with the exercise of reasonable diligence, known of its existence. Additionally, as discussed earlier, recognized hazards include those which are detectable either by the human senses or by use of special instrumentation. Therefore, hazard recognition can be established by either objective or subjective means.

Did the Employer Recognize the Hazard?

Evidence indicating that an employer had personal knowledge of a hazardous condition's existence demonstrates that the hazard was recognized. This does not mean OSHA must prove that accidents have occurred to prove an employer knew of a hazardous condition. Evidence of employer recognition of a hazard may be found in written (or oral) statements made by the employer to employees at some point before an accident occurred. OSHA states that company memorandums, safety rules, operating manuals or operating procedures, or even collective bargaining unit (CBU) agreements may reveal the employer's awareness (knowledge) of a hazard.

Publicized accident reports may also demonstrate employer knowledge. In addition, employee complaints or grievances to supervisory personnel may establish recognition of a hazard (if the evidence shows that the complaints were not merely infrequent, off-hand comments). Furthermore, the employer's own corrective measures may serve as the basis for establishing employer recognition of the hazard. This is especially true if the employer did not adequately continue or maintain the corrective action, or if the corrective action did not afford significant protection to the employees.

Did the Industry Recognize the Hazard?

A hazard may also be proven as *recognized* if it is considered to be common knowledge in the employer's industry. This means that even if the cited employer happens to be ignorant of the existence of a particular hazard or its ability to cause harm, the employer may still be in violation of the GDC if it is common knowledge in the employer's industry that such a hazard exists (i.e., is recognized). In this situation, the courts define "recognized hazard" as a condition known to be hazardous in the industry, even if not known by every employer. Industry recognition, then, is indicated by the general consensus of that industry's safety practitioners. Several means can be used to establish industry recognition of hazards. For instance, manufacturers or designers of equipment, machinery, or processes often issue warnings to industry and to product users about hazards associated with their equipment, products, or processes. Industry publications, advisory standards (such as those issued by ANSI), and state or local laws that address the hazard can be used as well.

But recognition of a hazard in another industry is not proof of recognition. Industry recognition must be in the same industry as that of the cited employer. Of course, when a particular hazard is clearly evident, the OSHRC may, without referring to industry practices or expert testimony, decide that it was recognized. At times, employers must use feasible abatement measures, even if they are not commonly utilized in the industry in question.

Was the Hazard Foreseeable?

GDC citations often result from an accident. In those cases, evidence must be available that the employer foresaw that the accident could occur. An employer's statement to that effect, or evidence indicating that the conditions which caused the accident were known to be hazardous in the employer's industry, can be used to establish *recognition. Foreseeability* is not established just because an accident occurred. Also, lack of evidence of similar accidents over the course of a number of years is evidence that the cited accident was, in fact, not foreseeable. The foreseeability element relates to the existence, *on the job or in the workplace,* of the hazard. In other words, the hazard must first be recognized (in the sense discussed earlier in this chapter) and, second, on the assumption that it is recognized, the employer must also have either actual or constructive knowledge of its presence on the job or in the workplace.

It should also be noted that the concept of foreseeability does not always apply to employee conduct on the job. According to the OSHRC, an employer's duty under the GDC must be one it can achieve. For instance, the GDC does not impose strict liability on employers for exposure to hazards resulting from worker's idiosyncratic, demented, or suicidal actions. In essence, an employer cannot be expected to control hazards created by employee actions which can not be reasonably foreseen. This means that employee conduct resulting in an accident is not reasonably foreseeable where the employer:

- Has established work rules designed to prevent the violation;
- Has adequately communicated the work rules to the employees;
- Has taken steps to discover violations of the rules; and
- Has effectively enforced the rules in the event of infractions.

These allowances acknowledge that nothing can insure compliance with safety practices at all times and that unexpected employee actions resulting in hazards cannot always be prevented. Hence, employers are only required to do that which it would be negligent for them not to do.

Could the Hazard Cause Death or Serious Physical Harm?

As previously stated, for OSHA to issue a GDC violation the hazard must not only be recognized, but must also be causing or likely to cause death or serious physical harm. The courts have not focused on the probability that an accident will happen, but on the likelihood that death or serious physical harm will result in the event of an accident. Therefore, the likelihood of an accident occurring need not be considered and the hazard need not actually have caused an injury or death for the "or are likely to cause" requirement of the GDC to be fulfilled, thereby substantiating a GDC citation.

How Feasible Is Hazard Abatement?

After establishing that a *recognized hazard* that is *causing or is likely to cause* death or serious physical harm existed, OSHA has to prove that a feasible and specific abatement method could have controlled the hazard. But the employer must realize that this proof will not appear in the initial citation presented to the employer. Whenever an employer contests a citation, the Department of Labor is obligated to file a more complete statement of the citation's charges with the OSHRC. This statement is known as the "Complaint." The Commission's procedural rules impose additional requirements on OSHA in GDC cases by providing that the complaint must "identify the alleged hazard and specify the feasible means by which the employer could have eliminated or materially reduced the alleged hazard" (29 CFR 2200.35). The term *feasible* has been defined by the OSHRC as economically and technologically capable of being done.

Technological Feasibility

A feasible means (i.e., technology) to abate the hazard must be proven to be capable of controlling the hazard that is the subject of the OSHA citation. If the hazard in question was the absence of an effective safety program, the agency must present evidence that the employer's safety program was insufficient and that specific alternative measures would be significantly safer if implemented. If OSHA claims that a toxic chemical is hazardous, then the agency has to prove that feasible technologies are available for reducing or eliminating the hazard.

Economic Feasibility

Even if an engineering or management control is technologically feasible, it may not be economically feasible. Feasible means capable of being done, executed or effected. The OSHAct did not intend for OSHA to engage in a cost-benefit analysis before promulgating a standard. However, the economic feasibility element of compliance may require a cost/benefit analysis be performed by the effected industry or by the cited employer to demonstrate that a particular hazard abatement strategy is not economically feasible.

The elements of proof discussed above are used whenever a GDC citation is contested. But there are other issues that the agency must prove in every case. For example, an employee's real or potential exposure to the hazard and its severity must be established. These elements and others common to all prosecutions under the OSHAct are discussed throughout this text.

APPLICATION OF THE GDC WITH EXISTING STANDARDS

If a standard regulating a specific hazard exists, an employer cannot also be held in violation of the GDC for that hazard. According to the OSHA Field Operations Manual (Section 190 of Chapter IV), OSHA inspectors shall not use Section 5(a)(1):

- "To cite hazardous conditions covered by an OSHA standard;
- "To impose a stricter requirement than that already required by an existing standard." For example, if a specific standard provides for a permissible exposure limit of 5 ppm, even though current data establishes that a 3 ppm level provides better protection, Section 5(a)(1) shall not be cited to require the 3 ppm level to be achieved.
- "To require an abatement method not set forth in a specific standard." A specific standard is one that refers to a particular hazard, such as a toxic substance, or deals with a specific operation, such as welding. If a toxic substance standard covers engineering control requirements but does not address medical surveillance, Section 5(a)(1) shall not be cited to require medical surveillance. In other words, just because a specific standard does not address what may be perceived by OSHA to be all the hazards or controls, it is not appropriate to use the GDC to fill in the gaps of the cited standard.
- "To enforce 'should' standards" on employers. If a standard or its predecessor (such as an ANSI standard) uses the word "should" instead of "shall or "must," neither the standard nor the GDC shall ordinarily be cited with respect to the hazard addressed by the "should" portion of the standard. The word "should" is normally construed to mean "recommended" or "suggested," whereas, the words "shall" and "must" generally mean mandatory action is required. The GDC can not be used to make a "should" recommendation into a "shall" or "must" requirement.

Employer Knowledge

In some instances, however, compliance with a specific OSHA standard does prevent an employer from being found guilty of a GDC infraction especially when the employer knows that just complying with a standard is not enough. Under the GDC, the employer must take any additional protective measures necessary to protect its workers against the specific hazard cited by OSHA if the employer knows the protections required by the standard are inadequate. In other words, employers will not be permitted to hide behind their compliance with a standard if they truly know that the standard is inadequate.

It must always be remembered that OSHA standards prescribe the minimum required actions to ensure a safe and healthful work environment. Intelligent engineering and effective management often dictate that these requirements be exceeded to a level sufficient to properly protect the safety and health of the worker. There is no record to show that OSHA ever cited an employer for being safer than a particular standard requires. The courts have, however, established *that employer knowledge* is the key to a GDC violation. If an employer knows that compliance with a specific OSHA standard will not protect its workers from a hazard, then the company will not satisfy the GDC obligation no matter how faithfully it observes the standard.

Chapter 5

PROPER RECORDKEEPING AND THE COMPLIANCE PROCESS

OVERVIEW

The requirement to make and maintain certain records related to occupational safety and health has been in place since the enactment of the OSHAct in 1970. Congress realized that it would be necessary to obtain such information to statistically determine the need for new rulemaking based on trends in accident cause data. To accomplish this, the Secretary of Labor and the Secretary of Health and Human Services were both granted the authority to require employers to make records of employee injury and illness data, to maintain those records for specific periods of time, and to make the records available to the Secretaries upon request.

During the first decade of OSHA, these requirements, although mandatory, were not strictly enforced on a uniform basis. Even when employers were found to be in noncompliance with the recordkeeping requirements, OSHA could fine them only once for the violation regardless of whether there was one failure to comply or many. It was still a violation of *one* standard and, hence, only *one* citation would be issued. During the Reagan Administration, however, OSHA enacted its *egregious policy,* which allows the agency to cite each individual's failure to comply as a separate violation, even if the same standard was at issue in each case. Therefore, employers could now face multimillion dollar fines based on recordkeeping violations alone. OSHA has extended this policy to all aspects of non-compliance. For example, if 20 grinding wheels in a machine shop do not meet the requirements of 29 CFR 1910.215(b)(9) for exposure adjustment of the safety guard, then the employer would be cited 20 times for the same violation.

OSHA requires employers to collect and maintain specific data related to employee injuries, illnesses, and fatalities. To facilitate this requirement, OSHA provides a log and summary form (the OSHA 200) which can be used by employers. Supplementary data, which is generally more specific than that which is entered on the OSHA 200, is also required. This information is placed on the OSHA 101 or similar form. The information required by the OSHA 200 log must be recorded within 6 days of any occurrence. The form for the previous year must be signed and posted annually in an area where all employees can see it for the period February 1 through March 1.

OSHA, employees, and any designated representatives of employees have a right to access, view, and copy (one cost-free copy) the OSHA 200, the OSHA 101, and all other exposure, monitoring, and medical records that pertain to their individual work duties.

Employers must maintain the OSHA 200 and supplemental information for a period of 5 years (29 CFR 1904). For medical and exposure records (e.g., air monitoring reports, material safety data sheets, audiometric tests, etc.), the employer must maintain copies for a period of 30 years beyond the length of an employee's employment.

INTRODUCTION

When Congress passed the OSHAct in 1970, it knew it would be necessary to gather statistics on occupational accidents, injuries, illnesses, and fatalities. This need was especially a concern in the area of occupational exposure to chemical substances. Congress realized that literally thousands of new chemicals were being introduced into the workplace each year. But very little information was available about the effects of occupational exposure to these chemicals. The Secretary of Labor (specifically OSHA) and the Secretary of Health and Human Services (DHHS) were empowered to ensure employers compiled statistical health and safety information and to require employers to prepare and maintain records.

At the Department of Labor, this authority has resulted in numerous recordkeeping requirements pertaining to occupational safety and health. The majority of these requirements are established at 29 CFR 1904 (Recording and Reporting of Occupational Injuries and Illnesses).

However, it should be noted that numerous other specific standards, such as the Hazard Communication Standard (29 CFR 1910.1200) and the Occupational Exposure to Bloodborne Pathogens Standard (29 CFR 1910.1030), for example, contain additional more specific recordkeeping provisions that must also be considered.

Whenever OSHA promulgates a new standard, the agency can require employers to monitor employees' exposure to hazards. In addition, OSHA can require medical examinations and tests for employees exposed to hazardous health conditions. These tests are to be provided at the employer's expense. As a result of these requirements, a database of information is compiled on the effects of hazard exposure.

Both the Secretary of Labor and the Secretary of Health and Human Services are authorized to issue rules requiring employers to make, keep, preserve, and make available (to either Secretary) records of their actions under the OSHAct related to occupational accidents and illnesses. Each agency is also required to issue regulations stipulating health monitoring that may be necessary to ensure compliance with a given standard (noise exposure, for example).

DEFINITION OF TERMS

To fulfill the objectives of Congress, the Act requires both OSHA and the HHS to publish regulations which require employers to report and record work related deaths, injuries, and illnesses. Recorded annually on OSHA Form 200 (Log and Summary of Occupational Injuries and Illnesses), employers must provide details on occupational injuries, illnesses, and fatalities (See Figure 5-1 for a guide to recording cases under the OSHAct). But this requirement does not include minor injuries that require only first-aid treatment and which do not involve medical treatment, loss of consciousness, restriction of work motion, or transfer to another job.

Before continuing this discussion further, it is necessary to define several key terms which are essential to the understanding of OSHA recordkeeping requirements. It is not uncommon to find that misconceptions about OSHA's definition of these terms often lead to errors (and subsequent citations) in an employer's recordkeeping process.

Figure 5-1: OSHA's guide for determining whether injuries/illnesses are recordable under the OSHAct (source: OSHA, Washington, DC).

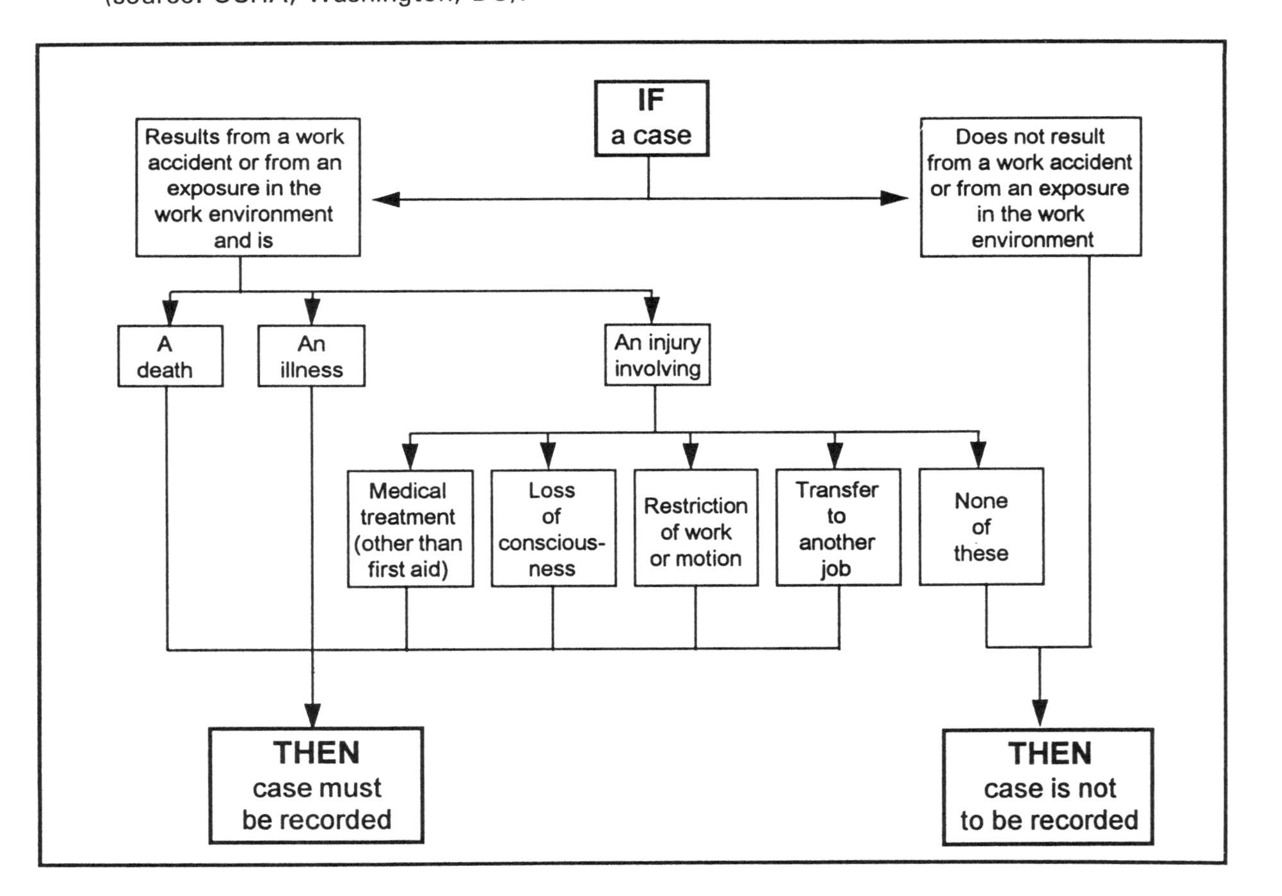

Occupational Injury

Any injury, such as a cut, fracture, sprain, amputation, etc., which results from a work accident or from an exposure involving a single incident in the work environment. In this definition, OSHA makes it a point to note that conditions resulting from animal bites, such as insect or snake bites or from one-time exposure to chemicals, are considered to be injuries for the purpose of recordkeeping.

Occupational Illness

Any abnormal condition or disorder, other than one resulting from an occupational injury, caused by exposure to environmental factors associated with employment. This includes acute and chronic illnesses or diseases which may be caused by inhalation, absorption, ingestion, or direct contact. For the purpose of classifying recordable illnesses, OSHA provides the following categories (and examples) of occupational illnesses and disorders. The examples provided are considered "typical" and are not a complete listing of all types of illnesses and disorders that should be counted in each category.

Occupational Skin Disease or Disorders. These include contact dermatitis, eczema, or rash caused by primary irritants and sensitizers or poisonous plants; oil acne; chrome ulcers; chemical burns or inflammation, and so on.

Dust Diseases of the Lungs (Pneumoconioses). Including silicosis, asbestosis and other asbestos-related diseases; coal worker's pneumoconiosis; byssinosis; siderosis, and other pneumoconioses.

Respiratory Conditions Due to Toxic Agents. These include pneumonitis; pharyngitis; rhinitis, or acute congestion due to exposure to chemicals, dusts, or fumes; farmer's lung, and others.

Poisoning (Systemic Effect of Toxic Materials). Poisoning by lead, mercury, cadmium, arsenic, or other metals; poisoning by carbon monoxide, hydrogen sulfide, or other gases; poisoning by benzol, carbon tetrachloride, or other organic solvents; poisoning by insecticide sprays such as parathion, lead arsenate; poisoning by other chemicals such as formaldehyde, plastics, and resins.

Disorders Due to Physical Agents (Other than Toxic Materials). Examples in this category include heatstroke, sunstroke, heat exhaustion, and other effects of environmental heat; freezing, frostbite; and effects of exposure to low temperature; caisson disease; effects of ionizing radiation (isotopes, X-rays, radium); effects of non-ionizing radiation (welding flash, ultraviolet rays, microwaves, sunburn).

Disorders Associated with Repeated Trauma. Such diseases include, but are not limited to, noise-induced hearing loss; synovitis; tenosynovitis, and bursitis; Raynaud's disease; carpal tunnel syndrome; and other conditions due to repeated motions, vibration, or pressure.

All Other Illnesses. Work-related illnesses not associated with any of the above listed categories but resulting from occupational exposure. These include anthrax; brucellosis; infectious hepatitis; malignant and benign tumors; food poisoning; histoplasmosis; coccidiodomycosis; HIV; and other viruses contracted in the work environment.

Medical Treatment

Includes treatment (other than first-aid) that is administered by a physician or registered professional personnel under the standing orders of a physician (e.g., nurse). OSHA is specific in stating that medical treatment does NOT include first-aid treatment (one-time treatment and subsequent observation of minor scratches, cuts, burns, splinters, and so forth, which do not ordinarily require medical care) even though provided by a physician or registered professional personnel.

Establishment

For the purpose of maintaining records, an establishment is considered a single physical location where business is conducted or where services or industrial operations are performed (for example: a factory, mill, store, hotel, restaurant, movie theater, farm, ranch, bank, sales office, warehouse, or central administrative office). Where distinctly separate activities are performed at a single physical location, such as construction activities operated from the same physical location as a lumber yard, each activity shall be treated as a separate establishment.

For firms engaged in activities which may be physically dispersed or spread-out, such as agriculture, construction, transportation, communications, and electric, gas, and sanitary services, OSHA requires records only be maintained at a place to which employees report each day. Records for those employees who do not primarily report to work at a single establishment (traveling sales representatives, technicians, engineers, and the like) shall be maintained at the location from which they are paid or the base from which personnel operate.

Work Environment

The physical location, equipment, materials processed or used, and the kinds of operations performed in the course of an employee's work, whether on or off the employer's premise, comprise the employee's work environment.

Lost Workday—Away from Work

A day on which the employee would have worked but could not because of occupational injury or illness. This does not include the day of the injury or onset of illness or any days on which the employee would not have worked anyway (such as a weekend or holiday).

Lost Workday—Restricted Work Activity

A day on which, because of injury or illness, the employee:

- Was assigned to another job on a temporary basis; or
- Worked at a permanent job less than full time; or
- Worked at a permanently assigned job but could not perform all duties normally connected with that job.

This does not include the day of the injury or onset of illness or any days on which the employee would not have worked anyway (such as a weekend or holiday).

The regulations do not provide clear definitions on several key elements of the recordkeeping requirements. For example, just exactly *who* is an "employee" and *what* constitutes a "work related occurrence" is somewhat ambiguous. Indeed, for recordkeeping purposes, what is an "injury" and what is an "illness" also is not clearly defined. The following definitions have been developed based upon interpretation provided by OSHA over the past several years.

Employer

The person authorized to direct and control the activities of another is the employer and the person taking the direction is the employee. Also, the person who supervises the employee every day is usually considered the employer. For recordkeeping purposes, temporary and part-time workers may be considered "employees".

Work Related Occurrence

OSHA considers an injury or illness occurring at an employer's workplace, to be *work related*, unless the employer proves otherwise. But there are some problems of interpretation which have developed under this approach. For example, injuries occurring in a restroom , hallway, stairwell, or lunch room, are considered work related under Bureau of Labor Statistics (BLS) guidelines. But, injuries occurring in parking lots, gymnasiums, and other such facilities are not ordinarily considered to be work-related injuries. To complicate this thinking further, injuries occurring in such locations may, in fact, be deemed occupational if the injured employee's job placed him/her in a parking lot (e.g., a parking attendant), or in a gym (e.g., a physical therapist), or at a ball field (a professional umpire, grounds keeper, etc.). Figure 5-2 provides some clarification on recording injuries and/or illnesses which may have occurred off the employer's premises but may still be questioned as to whether or not these incidents are to be considered occupational under the Act.

Figure 5-2: OSHA's guide for establishing the work relationship of injuries or illnesses that occur off the employer's premises (source: OSHA, Washington, DC).

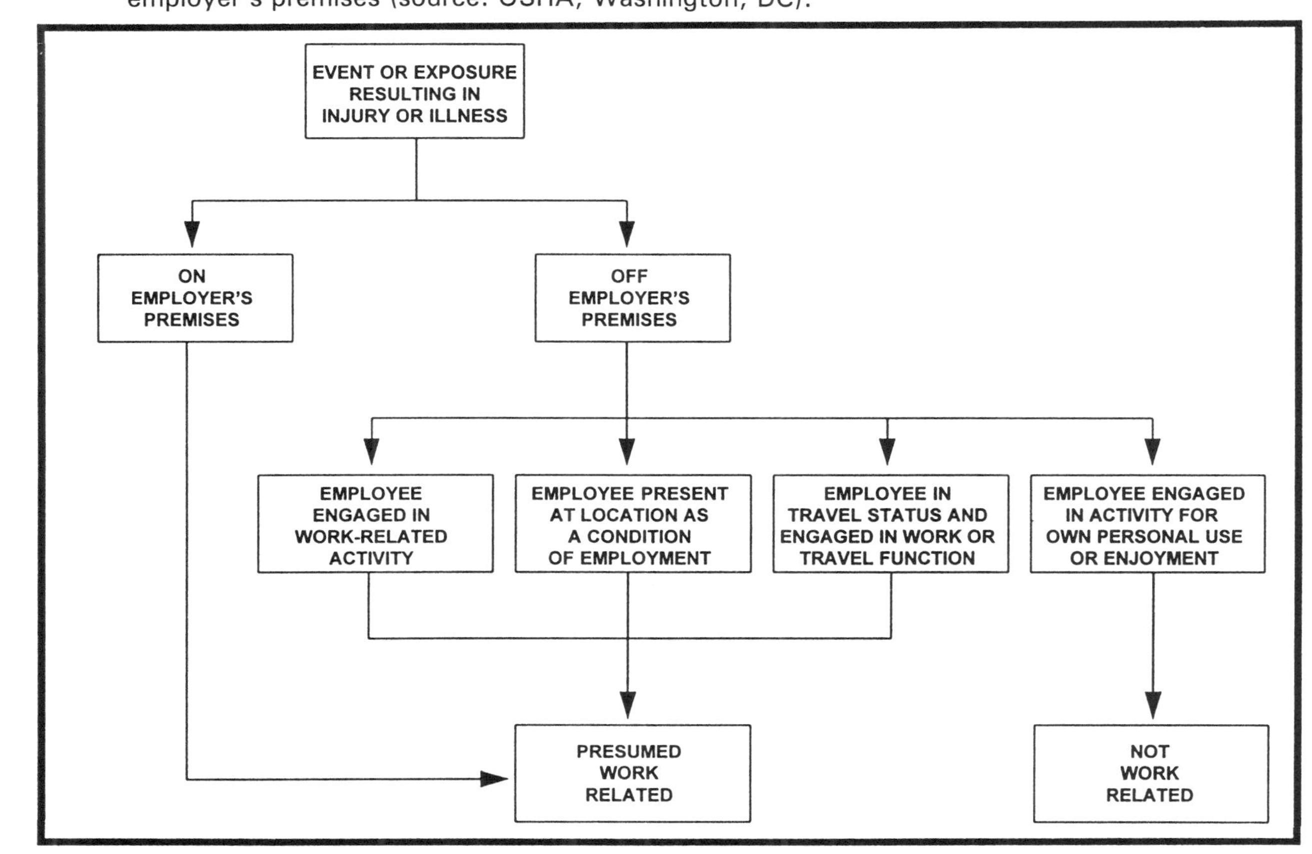

Characterizing and Recording Illnesses

The terms *occupational injury* and *occupational illness* have been defined above. However, some problems can arise when attempting to apply these definitions to actual occurrences. An incident only requiring first-aid is not considered recordable. Specifically, BLS guidelines stipulate that minor scratches, burns, splinters, and so forth, which do not ordinarily require medical care, even if performed by a physician, are first-aid injuries only and are not recordable. But, OSHA's Bloodborne Pathogens Standard (29 CFR 1910.1030) requires that a needle stick be recorded as an occupational *injury*. Even though the ordinary and natural course of action for a needle stick injury is to provide first-aid (clean the wound, apply antiseptic, protect with a bandage, etc.), OSHA wants this type of injury recorded. This is obviously in case the employee, as a result of the needle stick, develops Hepatitis or some other illness transmitted by a bloodborne pathogen. But the disease is an *illness* which should be recorded that way if it occurs and is work related.

The confusion in all this is that OSHA concentrates on the *event* as the cause of the illness and requires that event to be recorded as a work related *injury*. If an unfortunate illness does eventually develop, the question arises: should the employer record it again and have two entries for the same employee as a result of a single incident, but categorized two different ways? This type of situation reflects the flawed logic of recordkeeping. It is an example used here to highlight an imperfection in the system and demonstrate how the recording of injuries and illnesses continues to fall short of the original intent of Congress.

The Intent and Purpose of Recordkeeping Requirements

As mandated by Congress, OSHA requires employers to report fatalities, accidents resulting in the hospitalization of five or more employees (29 CFR 1904), and injuries and illnesses, with the BLS collecting the data. In issuing regulations related to exposure to toxic substances (such as the asbestos, lead, and cotton dust standards), OSHA usually includes provisions mandating the creation and maintenance of medical records regarding exposures. To facilitate the process, the BLS published guidelines for employers to follow in recording injury and illness data.

Recordkeeping as an Enforcement Tool

However, in terms of enforcement, these requirements were largely ignored by OSHA until the late 1980s. Until that time, the agency just did not place very much emphasis on these recordkeeping requirements. In fact, OSHA has used the recordkeeping requirements to determine which establishments to inspect. From the early 1980s until the practice was discontinued during the Bush years, an employer whose accident/injury and illness rates were below the national average could avoid a general OSHA inspection on that basis alone. The Reagan OSHA did realize that recordkeeping, or the lack thereof, could be used quite effectively to levy substantial fines under the *egregious* violations policy. This approach continues today.

Recordkeeping can also be used to help OSHA in its enforcement activities. A review of the regulations issued by the Bush OSHA shows that they frequently included provisions requiring employers to generate written compliance programs customized for their workplace. For example, the Lockout/Tagout standard, 29 CFR 1910.147 requires employers to prepare written energy control programs. The standards also required employers to prepare and maintain employee training records (example 29 CFR 1910.1030, Bloodborne Pathogens) and to inspect their own workplace for compliance. The records must also be made available to OSHA compliance officers, facilitating the agency's ability to detect non-compliance with the standards.

Egregious Violations

Prior to the egregious violation policy, an employer could be cited *once* for a recordkeeping violation regardless if there were one, twenty, or even two hundred or more alleged failures to record. Under the new policy, however, OSHA could impose a fine *for each* unreported recordable injury or illness. Additionally, under this policy, such violations were considered *willful,* punishable by maximum penalties of up to $10,000 per violation. In November of 1990, in an action that became known as the *seven-fold increase,* Congress amended the OSHAct, establishing a minimum penalty of $5000 and increasing maximum penalties to $70,000 per willful violation. OSHA has used this policy since that time to impose multimillion-dollar fines, primarily based on alleged recordkeeping violations. Virtually any violation of any standard can be subject to the egregious policy.

Congressional Mandate vs. Actual Practice

As of this writing, the Clinton Administration has not been able to accurately define its intent for OSHA's future. OSHA reform is widely discussed, and long-awaited rulemaking such as the Ergonomics Standard has a questionable future at best. However, it is probably safe to assume that if OSHA survives intact, its approach to recordkeeping will most likely survive with it and employers will still be responsible for recording and tracking employee accident, injury, illness, and fatality data.

It should be noted that none of the activities discussed thus far do anything to serve the Congressional mandate to obtain data to identify the sources of injuries and illnesses. Instead, the data details *how many* such incidents are occurring, but nothing about *what* causes them. Any subsequent rulemaking that results from these statistics is unlikely to be based on the actual cause of the accidents and illnesses they hope to control.

This state of affairs helps explain why many OSHA standards can be the result of efforts by special interest groups rather than actual need for regulation. For example, the impetus for regulation of occupational exposure to bloodborne pathogens originated with two employee unions. Upon petition, OSHA agreed to conduct a rulemaking and was proceeding in normal fashion when Congress intervened following pressures from the powerful labor unions. On 1 November 1991, Congress ordered OSHA to promulgate a final rule by 1 December 1991. If OSHA failed to do so, the proposed rule would go into effect.

Such Congressional interference in the normal rulemaking process, based solely upon the preference of a special interest group, is an all too common occurrence in our system of government. With regard to OSHA recordkeeping, it demonstrates how standards almost never result from statistical data, as Congress originally intended.

EMPLOYER RECORDKEEPING AND REPORTING REQUIREMENTS

Even though OSHA established injury and illness recordkeeping and reporting requirements early in its history, the requirements have never changed significantly, with one recent exception. In April of 1994, OSHA changed a key element of its reporting requirements. From the beginning, a single fatality or the hospitalization of five or more employees from a single incident required either oral or written notification to OSHA within 48 hours of the occurrence.

Reporting Deadlines

Under the amended rule (59 FR 15600, April 1, 1994), employers must now provide oral notification to OSHA within 8 hours following the work-related death of any employee or the in-patient hospitalization of three or more employees as a result of a work-related incident. The employer may make this notification either by telephone or in person to the OSHA Area Office that is nearest to the site of the incident. OSHA stipulates that this requirement applies to any fatality or multiple hospitalization that may occur up to 30 days beyond the date of the incident. The only exception is if the employer does not learn of a reportable incident at the time of its occurrence, then the report shall be made within 8 hours of the time the employer first learns of the incident.

Report Contents

Each report shall include the following information:

- Establishment name;
- Location of incident;
- Time of the incident;
- Number of fatalities or hospitalized employees;
- Name and phone number of contact person;
- Brief description of the incident.

There have also been changes to the guidelines published by the BLS aimed at improving the information provided to employers regarding what the BLS considers recordable injuries and illnesses. Though BLS guidelines themselves were most recently revised in 1986, on 1 January 1990, the Secretary of Labor gave OSHA BLS's recordkeeping authority, including the power to establish recordable data requirements (BLS continued to tabulate, analyze, and publish the data). In 1992, OSHA began treating the existing BLS guidelines as part of the OSHAct. This was an important development since "guidelines" published by a non-regulatory agency do not normally have the force and effect of law, as do regulations, with which compliance is required.

In other words, guidelines are normally thought of as non-mandatory recommendations or suggestions on how to accomplish something. However, as OSHA continues to publish new guidelines as proposals and conducts informal rulemaking with respect to the proposals, the "guidelines" become regulations. And since OSHA does have the authority to interpret the intent of the final guidelines or rules, recordkeeping requirements, whatever they may be and however they may evolve, are still subject to final OSHA interpretation. The employer will still have to comply.

The Log and Summary (OSHA 200)

Requires most employers (with specific exceptions) to create and retain injury and illness records for every establishment they maintain, 29 CFR 1904.2 through 1904.6. OSHA requires that the employer provide a summary of recordable injuries and illnesses. To facilitate compliance with this requirement, OSHA makes available a printed "log and summary," known as the OSHA 200 Form. The OSHA 200 does not have to be used, as long as the required information is recorded properly.

The following information must be recorded:

- Case or file number (company-assigned for tracking);
- Date of injury or onset of illness;
- Employee's name;
- Employee's occupation or job title;
- Employee's department;
- Description of the injury or illness;
- Number of days away from work (if applicable);
- Number of days restricted work activity (if applicable);
- Date of death of fatalities (if applicable).

Regardless of what form is used, the employer must record the above listed information within *six days* following the injury or onset of illness. The log and summary must be maintained at the location wherein the injury or illness occurred. It should be noted that some instances of occupational illness may never be recorded because of the extended latency period of some diseases. For example, asbestosis, silicosis, and other occupational lung diseases do not manifest with symptoms for many years after exposure. Unfortunately, there is no accounting for these illnesses in current OSHA recordkeeping guidelines.

The Supplementary Record (OSHA 101)

In addition to the log and summary, OSHA requires each employer to maintain a *supplementary record* for each occupational injury or illness occurring at a given worksite. Once again, OSHA makes available a form, known as the OSHA 101, for use in facilitating compliance with this requirement. But its use is not mandatory, just so long as some type of form is used to record the required supplemental data.

This information includes:

- Identifying information for the injured or ill employee;
- The location where the injury or illness occurred;
- A description of the nature of the injury or illness;
- The date it occurred;
- The parts of the body affected.

Included in the supplementary log should be information on any hospitalization and the identity of the person filing the report. Once again, while OSHA does provide the Form 101 for recording the supplementary information, worker's compensation forms and/or other insurance forms (which typically require the same information) are perfectly acceptable for use in place of the Form 101.

The Annual Summary

Employers who make and maintain these records are also required to post an annual summary of their injury and illness experience for the previous year. During the period of 1 February through 1 March of each year, this summary must be posted in an area of the establishment that is frequented regularly by the employees for their viewing. This may be a break area, a personnel office, a time clock location, or similar area where other employee notices are normally posted. The information appearing on the OSHA 200 is used to prepare the summary. The person responsible for its accuracy must sign the form indicating that the information is "true and complete." Even if there were no recordable injuries or illnesses during the previous year, the summary must still be prepared, signed, and posted for the length of the designated period.

The OSHA Poster

Incidentally, OSHA also requires employers to display a poster advising employees of their rights under the OSHAct. This poster should be 8.5" x 14" inches in size (OSHA has issued citations when the poster was not of this specified size). OSHA recommends that the yearly summary log be posted in the same location as the OSHA poster but does not actually require this.

Record Retention

OSHA requires employers to maintain the injury and illness records for 5 years and to provide them to the agency upon request. As a matter of course, OSHA will normally ask to review the OSHA 200 log or its equivalent at the beginning of any inspection (see Chapter 8). Employees are also entitled to see the OSHA 200 data, in addition to other information related to occupational safety and health (see Chapter 7).

Exemptions to Recordkeeping Requirements

Not all employers are subject to OSHA's recordkeeping requirements. Employers with 10 or fewer employees at any one time in the prior year are usually exempt. Employers in *standard industry classification (SIC)* codes which are considered to be low hazard industries in terms of employee illness and injury are also exempt. Examples include retail stores (such as department stores and shoe stores), restaurants, car dealers, health clubs, banks, real estate and insurance companies.

However, it should be noted that if the BLS solicits such establishments to participate in statistical surveys, they are required to keep and maintain records. Also, exempt employers are not exempt from other OSHA requirements, including reporting accidents resulting in fatalities to the agency within 8 hours of the occurrence. Some states also impose occupational safety and health recordkeeping rules which all employers must comply with.

It is important to understand that employers who may be exempt from OSHA recordkeeping requirements are not unilaterally exempt from all other safety and health obligations. Small employers must still comply with OSHA regulations and, while not subject to programmed safety inspections (see Chapter 8), they may still be inspected as the result of an employee complaint or to determine compliance with OSHA health standards.

Other Agencies' Requirements

Unlike other OSHA requirements, employers who are subject to other federal agencies' safety and health standards must still make and maintain OSHA records. This is true of all employers except small employers and those subject to the requirements of the Mine Safety and Health Act of 1977. In the case of the latter, the Mine Safety and Health Administration (MSHA) has established its own recording and recordkeeping requirements. Employers subject to the jurisdiction of other federal agencies are advised to review their recordkeeping requirements to clarify whether the OSHA forms or similar forms must be used.

The Recordkeeping Requirements of Specific Standards

Exposure Monitoring Records

Whenever OSHA promulgates a new standard that regulates hazardous substances or physical agents and requires exposure monitoring, it must also issue new recordkeeping requirements under those standards. OSHA has included such requirements with every standard of this type since the asbestos rule.

This type of standard is generally found in Subpart Z of 29 CFR 1910 and begins with the asbestos standard (29 CFR 1910.1001). These include standards for occupational exposure to inorganic arsenic (29 CFR 1910.1018), inorganic lead (1910.1025), coke oven emissions (1910.1029), bloodborne pathogens (1910.1030), and cotton dust (1910.1043). There are, of course, numerous other toxic substances for which their respective standards specify recordkeeping requirements. These standards require employers to monitor their workplaces for these substances. Initial monitoring is used to determine whether the substances are present in concentrations exceeding established action levels.

Other exposure levels established in such standards include the *permissible exposure limit (PEL),* the *short-term exposure limit (STEL),* and the *ceiling level.* Employers whose employees are exposed to concentrations exceeding the action level (usually one half the PEL), are generally required to conduct periodic monitoring and record the results. Other compliance activities not related to recordkeeping are also usually required at this point (e.g., providing personal protective equipment, implementing written programs, etc.). Employees who may be potentially affected by exposure, or designated representatives (unions, attorneys, for example) have the right to observe these monitoring activities and to receive a copy of any reports which may result from such activities.

Medical Monitoring Records

The standards regulating toxic substances and harmful physical agents also include medical monitoring and recordkeeping requirements. Under the asbestos standard, for example, a physician must periodically examine exposed workers. The lead standard requires testing of employees' blood lead levels and the maintenance of records of results by the employer. In fact, each standard states how long the records must be kept. In general, records must be held by the employer for either 40 years, or the length of the affected employee's employment plus an additional 20 or 30 years (depending on the standard).

Other Categories of Recordkeeping for Standards

Many general, construction, and maritime safety standards also include recordkeeping requirements. For example, recordkeeping requirements for respirators are found at 29 CFR 1910.134. Such requirements must be complied with in addition to the recordkeeping requirements of specific toxic substances standards.

Other standards featuring recordkeeping requirements include those for fire extinguishers (29 CFR 1910.157 and 1910.160); cranes and derricks (1910.179 through 181, and 1926.550); and mechanical power press operations (1910.217). Each of these standards requires inspections and recordkeeping documenting the inspections. Power press employers must also report point of operation injuries to OSHA within 30 days. Other standards, such as those for telecommunications, powered platforms, and commercial diving operations, all specify some level of recordkeeping. Therefore, all standards that may be applicable to an employer's business activities should be consulted to determine the precise extent of the recordkeeping requirements they impose.

Written Compliance Plans

It should also be noted that some OSHA standards (the Hazard Communication Standard, 29 CFR 1910.1200), the Lockout/Tagout Standard (1910.147) and the Bloodborne Pathogen Standard (1910.1030), to name a few, require employers to prepare *compliance plans.* Employers covered by the Bloodborne Pathogens Standard, for example, must prepare a compliance plan identifying the employees who are exposed to these pathogens. The plan must also provide a detailed explanation of how compliance will be achieved. Training records are also required under this standard.

The lesson to be learned from all the diversity between the various standards and their individual recordkeeping requirements is that employers must never assume they are in compliance with OSHA recordkeeping requirements simply because they have filled out an OSHA 200 log. All standards which are applicable to operations under their control or direction must be carefully evaluated for any requirements to make and maintain records.

GRANTING ACCESS TO RECORDS

OSHA Access

OSHA will typically begin an inspection (as described in Chapter 8) with a request to review the employer's OSHA 200 log and any other records the employer is required to keep. The inspector may also review any particular compliance plans which apply to the subject workplace (e.g., Hazard Communication Plan). The purpose of this review is to determine where and what will be inspected in the employer's establishment. The law requires employers to grant OSHA access any time a request is made.

Employee Access

The Act also requires that OSHA ensure employers are providing their employees access to exposure and monitoring records. To accomplish this, OSHA promulgated 29 CFR 1910.20 (Employee Access to Medical and Exposure Records). In addition to the provisions stipulated in this regulation, OSHA typically includes employee access requirements in any new standards it issues.

OSHA requires employers to maintain exposure and employee medical records regardless of whether a specific standard exists. For example, 1910.20 requires employers who monitor for the presence of airborne quartz (i.e., silica) to maintain the records they generate, even though there is no current silica exposure standard requiring such records. The point is, if an employer makes record of *anything* related to a possible employee exposure, then 1910.20 requires those records to be maintained and to be made available to affected or potentially affected employees.

Employers are also required to *inform* employees of their right of access at least annually; to maintain exposure records for 30 years (or for the duration of employment plus 30 years); and to provide access to said records to employees and OSHA. Even material safety data sheets, under the hazard communication standard, are considered to be employee exposure records and are subject to the records retention rules. This right of access is granted to both current and former employees who may have been exposed to the hazardous conditions at some point in their employment. Upon request, the employer is required to provide free-of-charge (first request only) copies of exposure monitoring records affecting the employee, and the employee's medical records. Access is to be granted within a reasonable period, ideally within 15 days of the request.

Designated Representative Access

While employers must grant an employee the right to access information and records that pertain to their working conditions, employers must also grant this right to any *designated representative* of an employee as well. This means that labor unions, relatives, and attorneys must, under certain circumstances, be given employee exposure and medical records.

As collective bargaining representatives for employees, labor unions have access under both 29 CFR 1910.20 and the National Labor Relations Act (NLRA). While the right of access to exposure records is guaranteed, there are restrictions regarding employee medical records (the NLRA permits employee representatives access to medical records as long as information identifying other employees is first deleted or obscured and if they first obtain signed authorization from the employee). Section 1904.7 grants employees and their representatives access to employer's OSHA 200 logs and OSHA 101 or other acceptable forms.

Access to Medical Records

OSHA can also request access to exposure and medical records. The DHHS can also request employee medical records for use in research activities only. However, even though the government has a right of access, the employer also has a right of privacy which extends to the records it must maintain. The employer can insist OSHA obtain a subpoena or warrant before granting access. Even though the OSHRC has upheld warrantless intrusion into logs (stating that employers have no right of privacy where such records are concerned), the employer may still wish to defend the right to ensure privacy, especially if the records contain information that is not covered by the OSHAct, or is privileged.

Chapter 6

THE COMPLIANCE PROCESS

OVERVIEW

OSHA holds employers responsible for complying with the provisions of the OSHAct, OSHA standards, the general duty clause, and other specific regulations. While the agency does acknowledge that employers may not always be able to control the actions of employees who sometimes violate standards, employers must still prove that all possible actions were taken to prevent the violation.

The concept of *employer compliance* will therefore be explored in this chapter along with information to help the reader understand the meaning of compliance and OSHA's interpretations and expectations associated with compliance. Basic guidance to assist in the development of a *compliance plan* will also be discussed. Such a plan will not only help employers document their compliance efforts, it may also serve as proof of compliance during litigation that may arise subsequent to any citations for alleged safety violations. The establishment and enforcement of workplace safety rules will be covered as well. It is not enough to develop, implement, and communicate safety policy; the employer must also provide enforcement and disciplinary actions when the rules are violated. Finally, information on the responsibilities at *multi-employer worksites* and *special employment* situations will be provided.

While it is clear that employer responsibilities under the OSHAct can become quite complex depending upon the situation, it is also clear that OSHA will hold employers responsible for compliance in most instances. Employers must therefore ensure their understanding of the OSHA requirements that apply to their workplace and take the necessary measures to achieve compliance.

INTRODUCTION

It is extremely important to understand that employers must comply with the provisions of the OSHAct (as discussed in Chapter 2), OSHA standards (as discussed in Chapter 3), the general duty requirements (as discussed in Chapter 4), and regulations such as the recordkeeping requirements of 29 CFR 1904 as well as those of specific standards (as discussed in Chapter 5). OSHA knows that employers cannot always control employees who may violate standards. But, the employer has to substantiate that all feasible preventative measures were taken to avoid the infraction.

Addressing *employer compliance*, this chapter provides information to help understand the compliance process, develop a compliance plan, establish and enforce workplace safety rules and responsibilities at multi-employer worksites.

UNDERSTANDING COMPLIANCE

It has been established thus far that employers must *comply* with safety and health standards that apply to their establishments, workplaces, and employees. For *general industry,* these standards are published in the Code of Federal Regulations (CFR) at Title 29, Part 1910. Standards which only apply to specific industries are found elsewhere in Title 29 (e.g., 29 CFR 1926 contains standards applicable to the construction industry). Regardless of the industry, it is the employer's absolute responsibility to ascertain which, if any, of the standards apply to its operations and to comply with all applicable requirements of those standards.

Compliance Is Mandatory

To understand their compliance obligations, employers must first acknowledge that complying with OSHA regulations and standards is neither a selective nor a subjective process. Compliance is a simple matter of fact, and all employers who operate an enterprise engaged in interstate commerce must do so in accordance with the requirements for such conduct established in applicable OSHA regulations or as a general duty.

While the above stated premise certainly sounds feasible and reasonable, an examination of real-world application reveals a great misunderstanding of compliance requirements. Issues of interpretation, applicability, intent, fairness, and consistency often complicate the compliance process. Unfortunately, while such concerns are being addressed or ignored, the safety and health of the worker can continue to remain at risk until resolution. It is ironic that the very workers such regulations are supposed to benefit often become bystanders in the compliance process.

OSHA Publications

To facilitate the process, OSHA provides an array of helpful publications. These include documents, brochures, guidebooks, and other similar materials, usually provided free of charge. While each may differ in subject, content, and presentation, the focus is generally on clarification and compliance assistance. But even with the help of these OSHA publications, compliance can still become a complicated issue for employers, large and small.

Determining Applicability of Standards

Large employers often have a staff of safety and health professionals, attorneys, and advisors available to evaluate OSHA standards. They determine applicability and intent, and advise the corporation on the best course of action to achieve compliance. While this approach proves very successful in most cases, problems can still arise due to matters of misinterpretation and assumptions on applicability.

For example, when the Process Safety Management Standard (29 CFR 1910.119) became law, many employers determined that it did not apply to their type of operations. Such decisions were based upon detailed evaluation of the published requirements. However, in response to numerous queries from industry, OSHA has issued a large number of *interpretive letters* and statements of intent since the enactment of this standard. These have served to clarify the agency's position on a variety of issues not specifically addressed in the standard or its preamble. As a result, the original understanding of compliance requirements has changed for some employers.

Unfortunately, not all employers in every industry will see each letter of clarification published by OSHA. It is therefore recommended that employers ask their own questions and obtain answers to ensure their understanding of compliance requirements.

Selective Compliance Is Not an Option

Some employers may lack the staff and expertise to properly evaluate the applicability of a particular standard. Their choices are limited. They can attempt compliance based upon their understanding of the requirements. They can hire a professional safety and health consultant to help them through the compliance process. Or, they can choose to ignore the standard under the assumption that their operation is simply too small a concern to attract OSHA's interest. The latter is, of course, a very dangerous choice for any employer.

Following the enactment of the Bloodborne Pathogens Standard (29 CFR 1910.1030), for example, many small, private practice physicians and dentists decided not to comply with the requirements set forth in the standard. It was complicated, time consuming, and some thought too expensive to justify the outcome. Many were already practicing exposure control procedures anyway. It was nothing new. Compliance seemed more futile than practical and, at the very least, a waste of valuable time and resources. However, the new law required specific actions on the part of the employer and selective compliance was not an option. OSHA has inspected many of the employers who opted for noncompliance and the fines and penalties have been, and continue to be, quite severe.

Once again, making sense of or understanding compliance means understanding that it is not a matter of personal choice or preference. It is a requirement of law and must, therefore, be taken very seriously. Only if a new standard does not apply is compliance not required.

Intent and Purpose

Before any decision is made regarding compliance, the standard in question must first be carefully evaluated for intent and purpose. Once again, OSHA requires employers to determine which standards apply to their particular situations and which do not. Obviously, compliance with those that do is only possible once an employer acknowledges this applicability.

All employers, large or small, have this same general obligation to determine applicability. This has proven to be a very complicated task for a great number of employers since a majority of cases brought before the OSHRC for resolution are based, at least in part, on confusion over the applicability of a given standard. The typical scenario: OSHA determines a standard applies and fines an employer who is not in compliance. The employer will contest on the basis of their assumption that the standard did not apply to their operations.

While it would be impossible to cite each and every case of this type and its outcome, suffice it to say that the Commission will generally favor OSHA's position unless the employer can prove otherwise. This burden of proof is often difficult but certainly not impossible, as volumes of case law favoring the employer will attest. *Understanding* compliance is the key.

Developing a Compliance Plan

One method of assuring proper consideration and understanding of compliance issues is to develop a compliance plan.

Self-Audits

While it is true that OSHA does not *generally* require employers to workplace safety and health audits (some specific standards may require this), OSHA will issue citations and penalties for uncontrolled hazards found during inspections or other enforcement actions. OSHA compliance officers typically interpret standards liberally. That is, they will view the standards as *including instead of excluding* hazards they encounter. Therefore, it is obvious that employers who wish to avoid citations must be proactive, auditing their worksites to find hazards and eliminate them.

Required vs. Optimal Plans

A plan for achieving compliance with applicable standards (or the GDC) is one suggested course of action. Some specific standards, such as the Hazard Communication Standard, the Lockout/Tagout Standard, and the Bloodborne Pathogen Standard, currently require some type of compliance plan.

Extending this approach to *all* operations is an excellent way to ensure hazard recognition, elimination, reduction, or control. It is worth noting that the so-called OSHA Reform legislation introduced in Congress on 1 August 1991 contains provisions which would require OSHA to promulgate regulations mandating workplace safety plans. While the future of this legislation is uncertain, employers might wish to be proactive with their compliance approach.

Plan Implementation

But having a workplace safety plan in place is not enough. It must be enforced as well. It should be a living document in that it is constantly under review and revised as required. A word of caution is warranted here. While such plans and policies are certainly more advantageous than potentially detrimental, employers should still be aware that any plans and audits they develop are considered "evidence" of hazard recognition. Failure to implement these plans or eliminate hazards could be considered willful misconduct. This is not to suggest that employers should avoid documenting their compliance efforts. It simply means that documenting them is not enough to prove compliance. Employers must also *act* on their plans with clear evidence of implementation and enforcement. It is a simple statement of fact that one of the most effective defenses against OSHA citations is a workplace safety plan that has been effectively implemented and consistently enforced.

Elements of a Compliance Plan

While a compliance plan for any given organization will, of course, be tailored with specific information applicable to that company's operations, the essential elements of any compliance plan should include the following.

Figure 6-1: The essential elements of a compliance plan.

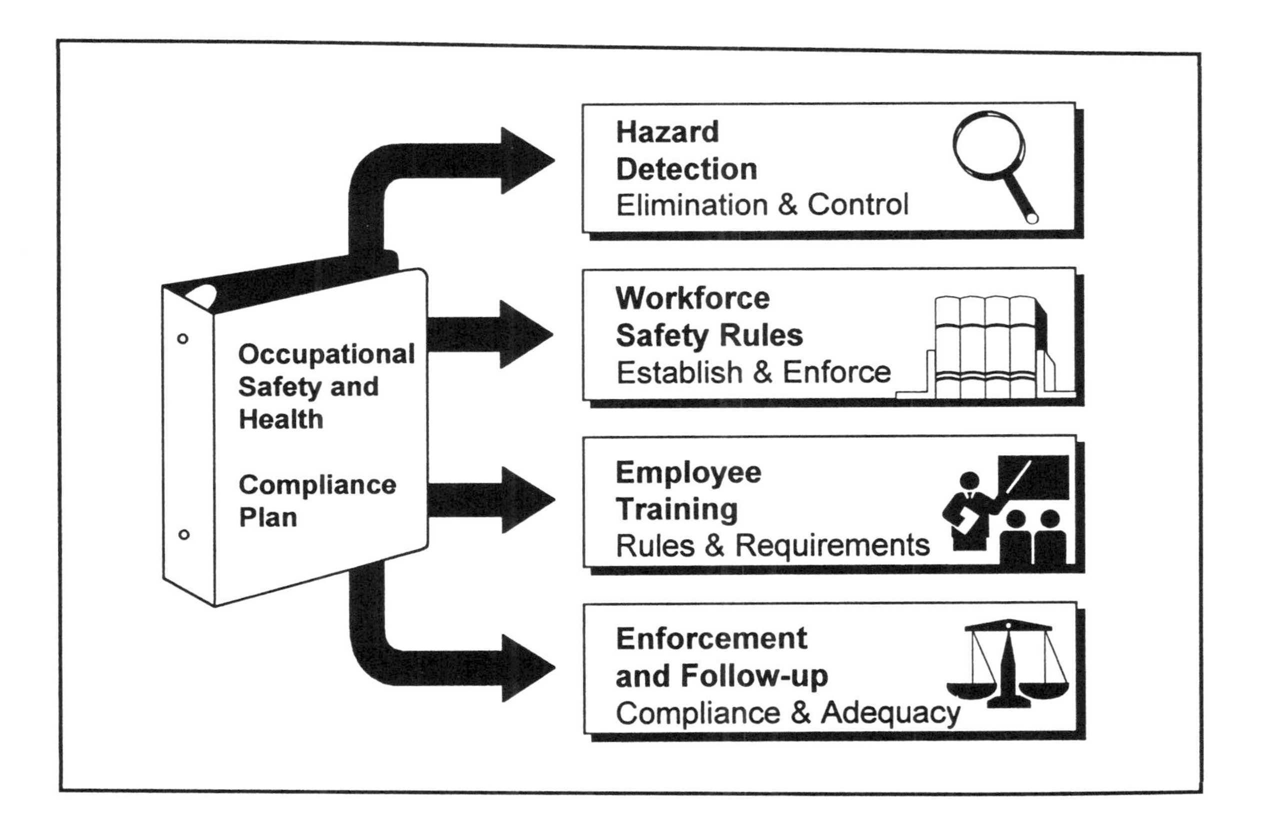

Detection and Elimination (or Control) of Hazards

In simple terms, employers are in the best position to detect hazards that exist in their places of employment. They should know their operations better than anyone else. When something is not right, the employer should be able to detect such irregularities, evaluate them, and determine the best course of action to return to normal operations. OSHA expects employers to provide a safe and healthy work environment. This requires a comprehensive approach to hazard recognition, evaluation, and elimination or control. The compliance plan should therefore require a scheduled, routine approach to hazard evaluation in the workplace.

Safety and health surveys, inspections, audits, or other such measures not only help the employer determine if new hazards exists, they also serve to document the adequacy (or inadequacy) of any control measures in place for existing hazards. By doing so, the employer can fulfill the compliance objective in two areas. First, new hazards to employees can be identified, evaluated, and eliminated or controlled in accordance with OSHA requirements. These requirements may be specific (such as confined space entry) or general in nature. But either way, the employer has a positive means in place to protect its employees from hazard exposure.

Second, the control of previously identified hazardous conditions can be examined and proven adequate through frequent evaluations. Documentation of findings can assist the employer in satisfying burden of proof requirements. Caution, however, should be advised when documenting compliance efforts. As stated earlier, audits that document hazardous exposures can be used by OSHA to demonstrate willful non-compliance *if no actions were taken to eliminate or control such exposures.*

The key, then, is to ensure that adequate, effective, and visible actions are taken once a workplace hazard has been identified. Such actions may be the installation of a mechanical guard, or the provision of explosion-proof electrical equipment, or any other hazard reduction technique that may be required.

Establishment and Enforcement of Workplace Safety and Health Rules

Providing mechanical or engineering abatement of workplace hazards is not enough either. All of the ventilation systems, machine guards, personal protective equipment, and the like are useless without enforcement of rules requiring their use. The employer must develop, implement, and enforce workplace rules which require the use and maintenance of any safety device or equipment. Rules must also be developed addressing expected employee conduct related to workplace safety and health.

It should be noted that *enforcement* is a key element of this equation. Even if the employer has safety and health rules which adequately address the subject(s) covered by OSHA standards, a citation is warranted if the rules are not enforced. Enforcement can take many forms, but at minimum, should consist of a uniformly administered disciplinary program which does not favor one segment of the employee population over another. The employer must audit compliance with its own safety and health rules and take progressive disciplinary action for noncompliance. Under a written disciplinary program, oral reprimands may be a first step, followed by written reprimands, time off, and finally dismissal for repeat violations.

The existence and enforcement of adequate workplace safety rules can also be a viable defense when an employer has been cited for an alleged OSHA violation resulting from, what the employer contends, was employee misconduct. By demonstrating that they had adequate rules and requirements in-place to address a particular hazard and that these rules were part of an overall compliance plan focusing on the elimination of recognized hazards, the employer may be able to show that it took all possible measures to prevent exposure. This approach may help solidify a claim that employee misconduct caused the exposure, not the employer's failure to provide a safe workplace. However, while the employee misconduct defense is very often used, it is not often successful.

Another issue that often appears in contested OSHA citations is OSHA's obligation to prove the employer had knowledge of violative conditions. In such cases, OSHA must prove that an employer knew, or with reasonable effort, could have discovered the violation. So OSHA requests records such as the company compliance plan, company safety audits and inspections, insurance reports, reports from consultants, safety committee reports, and any other documents indicating the employer knew an uncontrolled hazard existed in the workplace. Simply stated, if an employer has a published work rule or policy addressing a particular hazardous condition and an accident occurred anyway, OSHA might argue that the employer knew of the condition and failed to properly eliminate or control the hazard exposure.

In contested cases, there is a very important distinction of proof that can be a decisive factor. If the issue is *employer knowledge*, then OSHA has the burden of proof; a *responsible company employee's* knowledge of a hazard or violation is also considered to be knowledge of the company as well. If the issue is *employee misconduct,* then the burden belongs to the employer.

Does all this mean that employers would do well to avoid establishing workplace safety rules and policies? Of course not. This discussion simply demonstrates that establishing workplace safety and health rules means nothing unless they are uniformly and consistently enforced.

Means of Training Employees on Rules and Requirements

One way to ensure employees are aware of safety and health rules is to train them. Case law demonstrates the importance of this issue. It is absolutely vital that employers *communicate* their workplace safety policies, plans, and programs to their employees. It is not enough to simply create a safety manual, place it on a shelf, and require all employees to know and practice its content. For successful defense in contested cases, employees must also clearly understand the rules. Understanding comes through communication from the employer to the employee. OSHA takes the position that a program not properly communicated is considered a *paper program.* OSHA inspectors will often question employees during workplace inspections and investigations to determine whether or not the employees are receiving training from their employers. This is especially true when compliance with a particular standard requires specific employee training (e.g., hazard communication).

Therefore, the employer must put into place an effective safety training program which provides specific information on the company's safety rules and requirements. In addition, where specific OSHA standards require it, the employer must provide employees with training related to those standards. It is recommended that records be kept to document employee attendance and participation. Administration of a pre-test (i.e., at the beginning of the training) and a post-test (i.e., once the training is complete) is an excellent technique to document employee participation, demonstrate comprehension and understanding (or lack thereof), as well as indicate areas for improvement in the training itself. Such records can also help the employer show employee awareness and understanding in the event of a violation or accident. It is not uncommon to encounter an employee at the heart of an accident situation claiming a lack of training and that, as a result, the hazard or its accident potential could not be foreseen. By producing documents that not only show the employee's attendance in training but their comprehension as well, the employer can successfully defend against any subsequent citation that alleges poor communication and training.

Provisions for Enforcement and Follow-up Actions

As stated earlier, a compliance plan should be a living document, under constant revision whenever situations change or new rules are established. As a provision to the plan itself, there should be a self-audit process to ensure the adequacy of the compliance plan, to verify established methods are being followed, and to determine if disciplinary actions are adequate.

Also, employers may have established workplace safety and health rules, they may have adequate training and communication programs in place, and they may have follow-up actions and self-audit provisions in their plan. However, if there is no established method of enforcement, the plan is not complete. Furthermore, attempting to prove an employee misconduct defense will be even more difficult if the employer cannot show its enforcement policy. While case law demonstrates that OSHA has no real authority to take action against employees who violate rules, this is not true of employers. Enforcement can take many forms (see section 2, "Establish and Enforce Workplace Safety and Health Rules"). Whatever steps are taken, it is important that they be *consistent* from one case to the next, *well documented,* and clearly *communicated* to all employees that may be affected by such policies.

WORKPLACES WITH MULTIPLE EMPLOYERS

Earlier in this text (Chapter 2), a brief discussion on individual employer responsibilities at multi-employer worksites established that a multi-employer worksite can take many forms. A construction worksite with a general contractor and several specialty subcontractors is the most common type. In the general industry setting, the factory or plant owner can often have an outside contractor performing special work in plant or at some other work location. *Borrowed* employees or temporary employees supplied by an agency represent another scenario.

Contractor—Subcontractor Situations

At a multi-employer worksite, each employer is responsible for their own workers' safety and health. OSHA will typically cite the employer for a violation of standards whenever that employer has employees exposed to a hazard. In fact, it is common for OSHA to cite several employers for the same violation (such as poor housekeeping, for example). The basis for the citation is that the employees of each contractor (i.e., each employer) are exposed to the hazard. In theory, contractors or subcontractors not responsible for the hazards will force the responsible contractor to ensure abatement of the hazard.

Contractors and subcontractors should document and even complain about hazards not of their own creation to which their employees are exposed. In other words, the general contractor (or factory/plant owner) should be sent written notice of the hazards, requesting abatement by the appropriate party. In addition, if the general contractor holds meetings (either safety meetings specifically or construction progress meetings in general), the subcontractor should attempt to document its safety concerns in the meeting minutes, so that complaints are in *writing.* These documents provide the employer with substantial evidence to demonstrate to OSHA and/or OSHRC that it did everything possible to protect its employees at the multi-employer job site.

General Contractors Are Ultimately Responsible

Since OSHA will usually hold the general contractor responsible for the hazards at a construction site, general contractors should always try to eliminate all hazards at the job site. If a subcontractor is in fact creating the hazard, then the general contractor should exercise its administrative authority over the subcontractor to abate the hazardous condition(s). Any steps or measures taken should also be documented in case proof of action is required at some point in the future (i.e., following an OSHA citation). The employer who created the hazard will be held liable for violations of standards even if the employer's own employees are not exposed to the hazard. Therefore, the importance of properly documenting corrective actions and measures on the part of each employer cannot be overstated.

Loaning or Borrowing Employees

Careful consideration of several questions before entering into any agreement involving the loaning or borrowing of personnel may help an employer avoid citation. These questions include:

- Who is the responsible employer?
- Who has the power to direct and control the activities of the exposed employee?
- Who pays the employee's wages?
- What firm does the employee identify as his or her employer?
- Who has the authority to hire and fire?

Other Special Situations

Employers frequently rent equipment, such as cranes or other heavy equipment, from other employers. The lending employer may also provide equipment operators. Who is then the responsible employer when a violation of OSHA standards occurs involving the leased equipment? Case law shows that the lessee can be held liable if it ordered the operator of a crane, for example, to lift a load that was too heavy. But if the violation was the operator's fault, the lender will be liable.

Chapter 7

EMPLOYEES' RIGHTS AND OBLIGATIONS

OVERVIEW

At the time of its enactment, the primary intent of the OSHAct was to ensure the safety of the worker. Congress knew that employees would be an important element in this process and granted them certain rights under the Act. This chapter will focus on these employee rights and provide a brief discussion on the nature of these rights and their meaning to both the employer and the employee.

Employee rights under the OSHAct include the *right to complain* about safety and health conditions in the workplace which affect them. Complaints can be formal (written and signed by the complainer) or informal ("nonformal"), which may be anonymous phone calls, unsigned letters, and the like. For formal complaints, OSHA's required response is very specific: letters to the employer and, perhaps, an inspection of the location in question by an OSHA compliance official. For informal complaints, OSHA's response may be very different. Depending upon the nature of the informal complaint, the agency may simply send a copy to the employer with a cover letter asking them to review the situation and report back to OSHA within a specified period of time (usually 5 to 15 days). Or, the agency may send an inspector to investigate, although this is not likely in the case of the average informal complaint.

Employees also have a *right of protection* against discriminatory and/or retaliatory acts for exercising their right to complain. In fact, this right of protection extends to every employee right that is exercised under the Act. The Act even authorizes OSHA to obtain injunctive and back pay relief for employees who are discriminated against for exercising their protected rights under the Act.

The OSHAct also provides the right of an employee or their designated representative to *accompany OSHA inspectors* during their walk-around inspection of plants, factories, construction sites, and other workplaces. Employees can be interviewed by the inspector, with a guarantee of anonymity, without having any company management official present. In addition, employees or their representatives are given *limited contest rights.* Specifically, the employee or representative can contest the scheduled abatement date for correcting the violations if they feel it to be unreasonable. Incidentally, the Act describes an *employee representative* as a union official or labor organization, an attorney for employees, or "any other person acting in a bona-fide representative capacity of the affected employee(s)." The latter category includes, for example, family members or even an elected official.

The rights listed above were placed within the body of the OSHAct so that employees could participate in the process of ensuring a safe and health workplace. As case law developed, employees were also given the *right to remove themselves* from dangerous conditions under criteria specified by the courts. In essence, this right allows employees to refuse to work in any area or on any task that they truly and in good faith believe is dangerous to their safety and health.

Employers should be aware that such provisions exist, since a violation of any of these rights can result in citations, fines, and penalties. Similarly, employers should know and understand *employee obligations* under the Act. It is important to note that ensuring safety and health in the workplace is not a one-sided affair. Employers do bear the majority of the responsibility here, but employees must also do their part. The rights discussed in this chapter provide the employee a certain degree of involvement. Such insight will not only help avoid any misconceptions related to employee conduct, it will also assist in any enforcement policies the employer may develop.

INTRODUCTION

From the onset of the OSHAct, Congress insisted that workers accept part of the responsibility for improving workplace safety and health. For example, the *worker rights* provisions in the Act established the right of employees to complain about safety and health conditions and the right of employee representatives to accompany OSHA compliance officers on workplace walk-around inspections.

Additionally, employees are given the right to remove themselves from the workplace when hazardous conditions specified by the courts exist.

This chapter will explore the nature and extent of employee rights, as shown in Figure 7-1, and their obligations under the OSHAct. Employers should be aware that such provisions exist since a violation of any of these rights can result in citations, fines, and penalties. Similarly, employers should know and understand employee obligations under the Act. Such insight will not only help avoid any misconceptions related to employee conduct, it will also assist in any enforcement policies the employer may develop.

THE RIGHT TO COMPLAIN

The OSHAct guarantees employees the right to complain about conditions they feel violate OSHA regulations, or which present hazardous conditions or imminent danger to their health and safety. OSHA must keep the names of complainants confidential at the employees' request. OSHA deals with *formal* or *informal* complaints differently based on the nature of the alleged hazard(s).

Formal Complaints

29 CFR 1903.11 (Section 8 (f) of the OSHAct) established the requirements for formal complaints. Such complaints must be in writing and signed by one or more employees or designated employee representatives. The complaint must describe the hazard or imminent danger. This description must provide enough information to allow OSHA to review the allegations. In response, OSHA must conduct a site inspection as soon as possible.

Employees whose jobs expose them to the alleged hazard(s) can file formal complaints. It is important to note that the alleged hazard(s) can be located beyond their employer's establishment and can even be outside their employer's control. For example, a worker at a multi-employer construction site (see Chapter 6), may complain if he or she is exposed to a hazard from an adjoining facility.

Figure 7-1: Employee rights under the OSHAct.

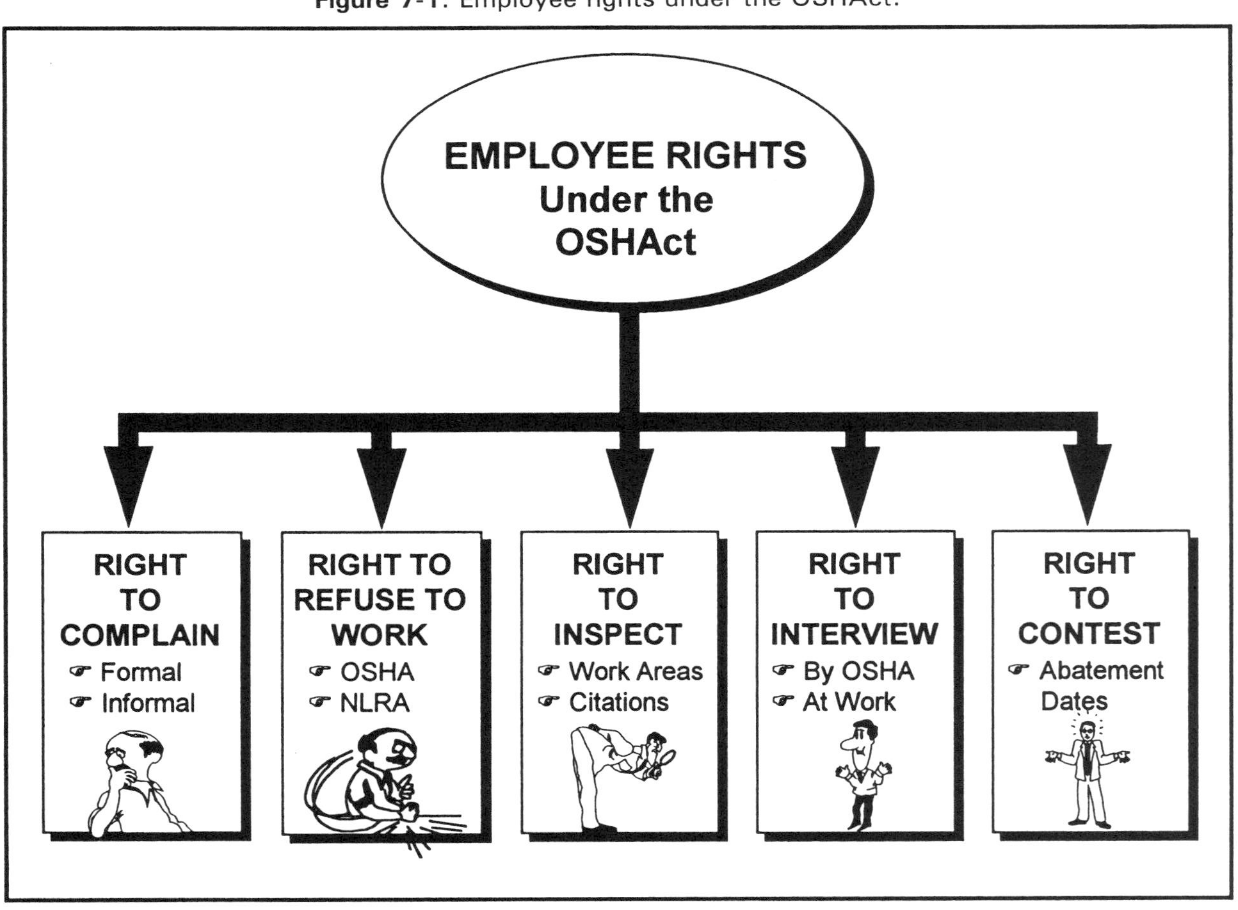

Complaints must include details of conditions believed to be hazardous, but the employee is not expected to know or include the exact title or paragraph number of the relevant standards.

Informal Complaints

Anyone, even a competitor, can file informal, or "nonformal" complaints. Informal complaints can be made over the telephone or in writing (unsigned written employee complaints). It is not uncommon for employees to file informal complaints. Their reasons for doing so vary, but the majority take such action because they are unaware of OSHA's obligation to protect their confidentiality.

Fear of retribution or retaliation from their employer, their immediate manager or supervisor, a union official, or even a fellow employee may also cause an employee to make an informal complaint. Hence, the OSHAct requires that OSHA hold the complainant's name in the strictest confidence whenever requested. The only exceptions to this rule are:

- When prior written statements signed by the employee exist, thereby invalidating his/her anonymity; or
- When the complainant is called to appear and provide testimony at a trial.

Incidentally, when employee written and signed statements do exist, they must normally be made available to the employer for use during cross-examination prior to trial (i.e., during discovery).

However, employers often know or suspect who the complaining employees are. This reality is a major reason for the protection provided employees under Section 11(c) of the Act. In essence, Section 11(c) prohibits retaliation or discrimination by employers against any employee who exercises her rights under the Act. This will be discussed in greater detail later in this chapter.

OSHA's Response

While OSHA is certainly obligated to listen to and accept informal complaints, it should be noted that OSHA Area Directors will generally exercise discretion when responding to them. Area Directors are expected to order an inspection of the suspect location, facility, or job site if the informal complaint alleges a hazard of imminent danger.

First, the OSHA Area Director will send a written *Notice of Alleged Safety Violation* to the employer alerting them to the allegation(s) contained in the complaint. The letter details which standards cover the complaint and asks the employer to investigate and abate any non-compliant condition found. In most instances, the letter will include information designed to help the employer conduct the investigation, such as information on the types of hazard(s) in question (copies of the applicable standard(s),etc.).

In addition, standard operating procedure (SOP) within OSHA requires the Area Director to select every tenth Notice of Alleged Safety Violation for an inspection. The Area Director will send a copy of the employer's reply to the complainant. The Director will request that OSHA be advised if the employee's proposed corrective measures are not taken or if the employee believes he/she is being retaliated against by the employer. Although the Act itself does not specifically require it, OSHA will tell the employer to post a copy of the Area Director's letter along with any related correspondence in an area where employees can view it.

The employer is asked to reply to the letter in writing within a specified period of time. This can vary depending upon the nature of the alleged hazard but is typically between 30 and 90 days. It is important to note that under the current regulatory framework, OSHA cannot compel an employer to reply to such letters, although it would certainly be wise for them to do so.

A Word of Warning

Employers should also understand that the information they provide in response to the Area Director's letter can be used against them in contested cases. Moreover, admission by the employer that an alleged condition is fact (i.e., that a violation of a standard actually does exist) may lay the foundation for proof of a willful violation should the conditions recur. This means that employers should avoid any language in their response which confirms any employee allegations or affirms the existence of a violative condition. The employer's written response should be carefully worded and structured so as to prevent self-incrimination. For this reason alone, it is always prudent and certainly advisable for employers to have corporate counsel review responses to an Area Director's letter before it is sent to OSHA.

THE RIGHT TO REFUSE TO WORK

Although not stipulated in the OSHAct, employees have a protected right to refuse to work when imminent threats to safety or health are present. In 1973, 29 CFR 1977.12(b)(1)-(2) was promulgated, which specifically protects an employee from discipline or discharge for refusing to work when "confronted with a choice between not performing assigned tasks or subjecting himself to serious injury or death arising from the hazardous condition of the workplace."

Under this regulation, five specific criteria must exist before a work assignment refusal will be protected:

- The refusal must be made in good faith (there must be a perceived danger) and not for ulterior motives;
- No reasonable alternative can be available to the employee;
- Any reasonable person should believe the hazardous condition poses a genuine risk of death or bodily injury;
- "Regular statutory enforcement channels" (i.e., a request for imminent danger relief from OSHA) would take too long, continuing to expose the worker to the hazard;
- The employee has previously requested that the employer eliminate the dangerous condition.

Refusal to Work Under the National Labor Relations Act

Section 7 of the National Labor Relations Act (NLRA) provides a similar employee right of work refusal. Specifically, employees have the right to engage in *concerted activities* for the purpose of *mutual aid and protection*. The Supreme Court has declared that refusal to work due to perceived or actual workplace hazards must be a concerted activity. Later court opinions have clarified this position, stating that workers must provide objective proof that the conditions were truly believed to be hazardous.

The key to this approach, then, is that the activity must be "concerted," that two or more employees must act together (in concert with each other). The actions of an employee acting alone must be supported or approved of by other employees or the activity will not be considered concerted under the NLRA definition. However, if a single employee asserts rights under an existing collective bargaining agreement, that action would also be considered a concerted employee action towards a common purpose or goal.

Determining Jurisdiction

It was obvious early in the history of OSHA that the possibility for a work refusal action could exist for the same situation under two different sets of rules (i.e., under OSHA and under the NLRA). To prevent a duplication of efforts by the National Labor Relations Board (NLRB) and OSHA, the two agencies published a memorandum of understanding in the Federal Register (40 FR 26083) in 1975. Basically, the memorandum states:

- *Where a charge involving issues covered by Section 11(c) of the OSHAct has been filed with the General Counsel of the NLRB, and a complaint has also been filed with OSHA as to the same factual matters, the General Counsel will, absent the withdrawal of the matter, defer or dismiss the charge. The General Counsel will inform the charging party of its action in writing, and will send a copy of its letter to OSHA.*
- *Where a charge involving issues covered by Section 11(c) of the OSHAct has been filed with the NLRB, but no like complaint has been filed with OSHA, the General Counsel will notify the employee of his/her right to file such a complaint within 30 days to OSHA under Section 11(c).* Then, if the employee notifies the General Counsel that an OSHA complaint has been filed or, if the General Counsel is so informed by OSHA pursuant to consultations at the end of the 30-day period, the General Counsel will, absent the withdrawal of the matter, defer or dismiss the charge. The General Counsel will inform the charging party of its action in writing, and will send a copy of its letter to OSHA.

- If, after an employee has been advised by the General Counsel to file with OSHA and does not do so, or if they do file with OSHA and then withdraw their charge, then under the NLRA, the General Counsel will press charges involving OSHAct Section 11(c) issues.
- *Where a charge has been filed with the General Counsel which includes issues covered by both Section 11(c) of OSHAct and within the exclusive jurisdiction of the General Counsel, the General Counsel together with the Office of the Solicitor of Labor will consult to determine the appropriate course of action.*

EMPLOYEE'S RIGHTS DURING AN INSPECTION

The OSHAct states that an employee representative should accompany OSHA Compliance Safety and Health Officers (CSHO) during workplace inspections. In union shops, this usually means a shop steward or union president. When no designated employee representative exists (as may the case in a non-union shop, for instance), then the inspector must consult with a representative sample of the total employee population. These issues are also found in 29 CFR 1903.7, as amended. The OSHA Field Operations Manual (FOM) establishes the CSHO's options for this.

The FOM stipulates that employees or their representatives should be present during every phase of an inspection. This includes the opening conference, closing conference, informal discussions and consultations with the employer, and during the walk-around inspection.

If an employer should reject employee participation, the CSHO must advise the employer that such participation is the employee's right under the OSHAct. Any further resistance by the employer would be considered a refusal to permit the inspection or an attempt to obstruct it. This can result in the issuance of a warrant from the U.S. District Court, commanding the employer to comply with OSHA's requests as set forth in the warrant.

Whether or not employers had to pay the employee for the time spent accompanying the CSHO during an OSHA inspection was an issue from day one. OSHA's position was that any refusal to pay could be construed as discrimination for exercising employee rights under the OSHAct. However, the courts have ruled that any mandate by OSHA to require employers to pay for this time is considered an "impermissible exercise of legislative power by an administrative body." So it is currently not considered discrimination when an employer refuses to pay employees during these inspections. It should be noted, however, that current proposals to reform OSHA have included provisions to require employers to pay employee wages when they assist the CSHO during an OSHA inspection. This issue continues to cause debate on both sides.

Employee Representatives

According to the OSHA FOM, an *employee representative* should be:

- A representative of the "recognized or certified bargaining agent" (i.e., the labor union, if applicable);
- An employee member of a safety and health committee selected by the employees; or
- An individual employee selected by the employees.

The employee representative does not have to be an actual employee of the company. A representative may be a union official, consulting industrial hygienist, safety engineer, or "other experienced safety and health person" chosen by the employees (e.g., some unions may provide their own safety and health consultants as part of their services to their members).

In some situations, there may be multiple employee representatives at a worksite. This is especially true at many construction sites. In these circumstances, the CSHO will exercise discretion. For instance, as long as the CSHO feels that multiple employee representatives will benefit the inspection and not hinder it, then all representatives will be allowed to be present. However, it is more common for the inspector to meet with the appropriate representatives of each area when their respective areas are actually being inspected.

When OSHA Interviews Employees

CSHOs often interview employees during the inspection. Employees have the right to speak to inspectors and even disclose possible violations. The CSHO is instructed by the FOM to make the effort to conduct any such interviews privately, even if it is not specifically requested by the employee. Employee interviews almost always occur in the absence of the employer or any representative of the employer's management. Discussions between the employee and the CSHO are usually held in confidence. Tape-recording interviews is allowed, but conversations may not be recorded against the employee's wishes.

Further interviews can be arranged by the CSHO at a mutually convenient time if employees request them. For example, interviews can be held at the employee's home, at the OSHA Area Office, or at any appropriate private location. As a general practice, interviews are documented in writing, in the first person, and in plain language the employee can understand. The documented interview should reflect only the content of the discussion, with any changes coordinated with and initialed by the subject employee.

The Right to Contest Citation Provisions

The OSHAct grants employees the right to file a notice of contest if they believe the period of time established in a citation for a violation's abatement is "unreasonable." Such notice must be filed with OSHA for forwarding to the Occupational Safety and Health Review Commission (OSHRC). After holding a formal hearing under the Administrative Procedures Act, the OSHRC must issue an order that either affirms or modifies the time period within which abatement must be accomplished.

Employees have 15 working days from the date they (or their representatives) receive first notice of the citation's issuance to file their notice. Employers can not keep such notice from their employees in an effort to prevent contest or for any other reason. In fact, employers are required to prominently post a copy of the citation(s) at a place near where the alleged violation(s) occurred in the workplace. The citation(s) must remain posted for a period of not less than three days so that employees may have ample time for notice of these allegations. Posting on a Friday and removal on the following Monday does not necessarily meet the intent of this OSHA requirement and would certainly invalidate any good faith in compliance issues the employer may attempt to pursue. Also, the citation will always include a date by or a period within which abatement of the alleged violation(s) must be effected. Hence, it is not in the employer's best interests to keep the notice of alleged violations from their employees. If any such actions have been taken, it is a serious violation of an employee right and can result in the issuance of fines and penalties against the employer.

Any employee or employee representative can file a notice of contest, but the employees who generally have the right to file a notice of contest are those who *are affected by* the alleged violation(s).

Under current requirements, the scope of the employee notice of contest is effectively limited to litigation of the abatement period only. However, proposals for OSHA reform have provisions which expand this employee right beyond the singular issue of the abatement period. Some versions will grant the right for employees to contest virtually every aspect of the citation. While it is uncertain whether any of these changes will ever be made law, it is clear that the sponsors of these proposals, as well as the private and public interest groups lobbying for OSHA reform, believe that the employee deserves more rights with regard to contesting citations.

The Right to Participate in Hearings

The OSHAct also stipulates that the OSHRC is to provide affected employees or their representatives an opportunity to participate in any hearings regarding citations. The OSHRC has interpreted this to include employee intervention in employer contests. In fact, OSHRC rules require employers to advise their employees when a contest has been filed and that they (the employees) have a right to participate. Employees must first elect *party status* and be officially recognized by the courts as a party affected by the citation. This election must be made at least 10 days prior to the start of a scheduled hearing. Having party status allows affected employees to present evidence at the hearing and to even cross-examine OSHA and employer witnesses.

Over the years, the extent of employee participation in OSHRC proceedings has been questioned. The OSHRC, on several occasions, affirmed employees' right to participate in all Commission matters.

OSHA agreed with this position up to 1983. Since that time, some courts have agreed with OSHA's position that employees should only be allowed to address and participate in issues related to abatement dates, since this is the only area they can contest under the current OSHAct. Perhaps this disagreement is one reason why the OSHA reform proposals require employee involvement in all aspects of a citation.

Lastly, the OSHRC requires employers notify all affected employees of any proposed settlement agreements (almost all employer contests end in settlement agreements), regardless of whether or not employees have party status. Employees can only object to the abatement dates agreed upon by OSHA and the employer. But if the agency decides to withdraw a citation, employees have no recourse.

THE RIGHT TO FILE DISCRIMINATION COMPLAINTS

As previously discussed in this text, Section 11(c)(1) of the OSHAct prohibits any retribution, including termination, against employees for exercising their rights under the Act, including testifying in hearings. Furthermore, employees can file complaints with the Department of Labor if they feel discrimination has or is occurring in this regard.

OSHA regulations help define the scope of the Act more clearly in this regard. Specifically, discrimination against any employee for the following actions taken under or related to the OSHAct is prohibited:

- Filing a complaint;
- Instituting proceedings;
- Testifying in proceedings;
- Exercising rights on one's own behalf or on behalf of any other person protected by the Act;
- Refusing to perform unduly hazardous job tasks (see discussion on The Right to Refuse Work presented earlier in this chapter).

It should be noted that, in OSHA's opinion, the Act prohibits discrimination by anyone, including employment agencies, unions, and the employer. While OSHA often uses an *economic realities* test that weighs a number of factors to determine whether an employment relationship exists, the fact is that a formal employment relationship with the alleged discriminator does not need to exist. In addition, individual officials of an employer's company can be held liable for retaliatory firings or discrimination.

Protected Activities

Over the years, discrimination cases have been proven in nearly all types of *protected activity.* Protected employee activities include any conduct or action which furthers the OSHAct's goals. The Act protects against virtually all forms of discrimination, so any actions which fall short of termination (e.g., suspension, time off, demotion, pay or grade reductions, being passed over for promotion when all other factors are equal, etc.) may also be interpreted as discrimination.

What Judges Must Decide in Discrimination Cases

The question of whether discrimination against an employee has occurred is a matter for the courts. This means that the facts of the situation must bear witness to and support the alleged act of discrimination. Once discrimination has been determined, an action appropriate to the level of damage resulting from or arising out of the retaliation or discrimination is ordered by the court. For example, back pay with interest is almost always a remedy in cases where employees lost their jobs or suffered a loss or interruption of income as a result of the retaliatory acts of the employer. Reinstatement is another option.

While some courts have allowed private causes of action, the courts do not normally permit employees a private right of action against those who have discriminated against their rights under the OSHAct. Each case is, of course, different, and the courts will allow for this in their review of private action requests. However, under most instances, a party cannot be accused twice of the same action under two different approaches to the law.

Discrimination Complaint Procedure

A complaint can be filed by an employee or authorized representative with the OSHA Area Director responsible for either the area of the employee's residence or the place of employment within 30 days of the alleged act of discrimination. However, OSHA applies what is known as *equitable tolling* in consideration of those cases where the employer may have deceived the employee or withheld information from the employee regarding the reasons for termination or other act of discrimination. In other words, the employee must file within 30 days of their first becoming aware of the discriminatory actions.

Once the complaint has been received by OSHA, the agency will generally notify the complainant within 90 days of its decision as to whether or not discrimination has occurred. However, this time period is not mandatory and actions based on the complaint do not automatically end if findings are not released by then.

Chapter 8

MANAGING THE OSHA INSPECTION PROCESS

OVERVIEW

OSHA enforcement actions are key to ensuring compliance with the provisions of the OSHAct. Such activities are the only way the agency can truly know the status of employer compliance. Inspections are a primary element of the enforcement process. While OSHA is authorized to enter and inspect all workplaces covered by the Act, the agency generally inspects less than two percent of these sites in any given year. Because of limited resources and the fact that OSHA could never really visit all six million American work locations each year (even with unlimited resources this would still be a monumental undertaking), the agency has established a scheduled priority of inspection activity. From the highest to the lowest, these priorities are:

1. Workplace situations that present an imminent danger of causing death or serious injury are inspected first.
2. OSHA will respond after a catastrophic accident involving a fatality and/or the hospitalization of three or more employees.
3. OSHA will respond to formal employee complaints of alleged safety violations.
4. OSHA will regularly schedule inspections of high hazard industries.
5. OSHA will revisit a previously inspected facility to verify compliance and abatement practices are adequate.

Except in rare circumstances and then only with permission from the Secretary of Labor, it is illegal to give advance notice of an impending OSHA inspection. Those found guilty of doing so can be fined up to $1,000.00 and/or jailed for up to six months.

Above all else, an inspection is *an evidence-gathering mission.* Simply stated, OSHA will attempt to obtain any and all information that can be used to prove a violation exists. It is not a social gathering nor is it an opportunity to obtain free consultative services from OSHA. It is a very serious matter that deserves management attention and, at the very least, their concern.

The inspection process starts with the *opening conference* with the employer and employee representative. The next step is usually a *record review* by the inspector who will look for accident trends and any other information to help the inspection. A *walk-around* through the facility will allow the inspector to gain an overall understanding of the work area. A *closing conference* will recap everything that was learned and summarize the inspector's next course of action with respect to citation.

The are basically two types of inspections: *safety* and *health.* The process for both is nearly the same. Each involves employee interviews and each can result in citations. However, where necessary, the walk-around portion of a health inspection may also include the monitoring of health conditions at the worksite (such as air sampling, noise monitoring, radiation monitoring, and the like).

This chapter will focus on the many issues and concerns associated with the entire inspection process. From employer responsibilities and employee rights, to the often complicated issue of liability at multiple-employer worksites. In making sense of OSHA compliance, it is important for every employer to know and understand what to expect from an OSHA compliance officer during a worksite inspection. But it is equally important for the employer to know what the OSHA compliance officer expects from the employer during such an event. Mutual professionalism, cooperation, and understanding of each other's position and responsibilities are key elements to a successful relationship with the agency. The employer must never forget its rights during an inspection and should never answer questions it does not understand or provide unnecessary (and unsolicited) information to OSHA.

INTRODUCTION

OSHA is empowered to enter workplaces and perform safety and health inspections in order to ensure compliance with the OSHAct. These inspections are divided into two categories. The first, known as *programmed inspections,* are conducted as part of OSHA's regularly scheduled inspection process. The second, referred to as *unprogrammed inspections,* are performed in response to particular events that occur during the inspection year, such as catastrophes, fatal accidents, and employee complaints. *Compliance officers* (OSHA's term for inspectors) are typically safety professionals or industrial hygienists (health professionals). The primary duties and responsibilities of the compliance officers are to:

- Inspect work locations;
- Compile evidence of failure to comply whenever they find it;
- Recommend the issuance of citations and penalties by the Area Director, based upon the severity of the noncompliance situations that they have discovered;
- Testify on behalf of OSHA when employers contest their findings;
- Conduct special inspections to determine the position OSHA will take with regard to employer petitions to amend or extend abatement dates, as well as with respect to various types of applications made by employees.

Without a doubt, OSHA's primary methods for ensuring compliance with the OSHAct have been the inspection process and the agency's ability to enter work locations, either with or without a warrant, and levy fines and penalties for noncompliance to effect an economic and legal burden upon employers.

WARRANT REQUIREMENTS

OSHA's Authority

Section 8(a) of the OSHAct authorizes OSHA to:

- Enter without delay and at reasonable times any workplace or other location where work is performed; and

- Inspect and investigate during regular working hours and at other reasonable times, and within limits, any such place of employment and all pertinent conditions, structures, machines, apparatus, devices, equipment, and materials therein, and to question privately any employer, owner, operator, agent, or employee.

While the Act would seem to authorize warrantless inspections (and prior to the landmark *Marshall v. Barlow's Inc.* [436 U.S. 307] case in 1978 this was indeed OSHA's approach), the Supreme Court ruled that owners have the same Fourth Amendment rights as private homeowners. The Fourth Amendment to the United States Constitution prevents arbitrary invasion of privacy and security by government officials. Specifically, the court ruled that:

> The businessman, like the occupant of a residence, has a constitutional right to go about his business free from unreasonable official entries upon his private commercial property. The businessman, too, has that right placed in jeopardy if the decision to enter and inspect for violation of regulatory laws can be made and enforced by the inspector in the field without official authority evidenced by a warrant.

As a result of this now famous ruling, OSHA must obtain the employer's permission or a warrant before entering a location scheduled for inspection. The warrant application must specify why the particular business establishment is to be inspected. The application must be signed under oath by the Area Director or an official OSHA designee. It is signed by a judge from the District Court having jurisdiction. Therefore, the information contained in the warrant application must be specific enough to satisfy an impartial United States Judge that a violation of the law *probably* (not possibly) has occurred. The judge must make an independent evaluation of the asserted basis (i.e., the *probable cause*) or the inspection.

A warrant can be requested at the beginning or during an inspection. Time limitations and other restrictions on the scope of the inspection (i.e., usually restricted to the reasons and/or locations stated) are included on the warrant. In contrast, if an employer agrees to a warrantless inspection, there are typically no legally stated time, scope, or objective restrictions involved (unless the employer specifies such conditions at the time of consent and OSHA agrees to them).

The Probable Cause Requirement

As stated above, OSHA must prove that probable cause exists before a search warrant will be issued. There are two principal ways in which OSHA can establish probable cause.

Specific evidence OSHA has or knows of showing that violations may exist at the workplace targeted for inspection can establish probable cause. This test is most commonly used to perform unprogrammed inspections because such inspections are usually restricted to the specific issues inspiring them. Evidence of probable cause most often comes from:

- Written or oral complaints from employees;
- Referrals from another agency (such as the EPA);
- Newspaper articles or other significant media attention surrounding a specific event at a work location;
- The need to conduct a follow-up to a previous inspection.

Accident reports, compliance officer's observations of hazardous conditions, and statements on the use of toxic substances made by the company can be other sources of probable cause.

The agency can also prove that the proposed inspection is part of a neutral administration selection plan meeting legislative or administrative standards. This test is known as the *systematic plan* and is used to support OSHA's programmed inspection program. Under this probable cause test, OSHA's application for warrant must:

- Specify the plan used to select the subject workplace for inspection; and
- Prove that the plan was unbiased in choosing the workplace for inspection.

Inspection—Selection Plans

Three key features characterize OSHA's inspection-selection plans.

- The Industry Rank Report: A report ranking industries (e.g., agriculture, petroleum refining, transportation, etc.) according to their lost workday injury (LDWI) rate which is sent to each local Area Office.
- The Establishment List: A list that contains the names of workplaces in each OSHA Area Office's jurisdiction that belong to the industries that have been noted on the industry rank report.
- The Inspection Register: These registers are complied by the local Area Offices using the above two lists. It consists of the names of the workplaces targeted for inspection and the order in which they will be inspected. Inspections may be conducted in any order to maximize resources (including time, personnel, budgets, and so on).

In states with their own OSHA-approved occupational safety and health programs, the methods used to target workplaces for inspection varies from state to state.

How the Evidence Is Weighed

When a judge reviews OSHA's evidence supporting a warrant request, a determination must be made as to whether probable cause does, in fact, exist. To do this, the following four factors are considered:

- Specificity of the Information: The information provided must supply enough detail on the alleged hazardous condition for the judge to determine whether the OSHAct has been violated. But this is not as difficult as it may seem. In most instances, the compliance officer's affidavit supporting the warrant will suffice.
- Likelihood of Violation: A plausible reason to believe a violation may have occurred can demonstrate the need for an inspection.
- Corroboration by Other Sources: If other sources exist which support the information presented in the warrant, then they should be listed in the application.

- Staleness: In some cases, the alleged conditions may be of the type that are likely to disappear through the mere passage of time, making immediate inspection and verification critical to establishing the existence of a violative condition.

Because OSHA often does not know whether or not a violation has occurred, the agency generally does not have specify which regulations it suspects the employer has violated.

Consented Searches vs. Warrant Requirements

Companies must decide whether to grant permission for OSHA entry to their premises when requested or to insist on a warrant. The company attorney will almost always advise the latter, more as a matter of legal principle than one of practical application of sound safety and health compliance strategy. In fact, many companies have policies against warrantless searches, so inspectors are usually told by the company safety or management representative that company policy will not allow OSHA entry without a warrant. Once the local OSHA office is aware of this position, the compliance officer will likely have a warrant in hand prior to seeking permission to make any future inspections.

Many companies, however, permit warrantless inspections under most circumstances. These companies know that requiring an inspector to leave the premises, travel back to his/her office, request his/her superiors to pursue the warrant, only to return again later with the requested warrant actually affords very little protection to the company and serves no real purpose, except of course to satisfy the company attorneys, antagonize the inspector, and delay the inevitable inspection. The actual, real-world advantages of requiring the warrant do not typically outweigh the potential disadvantages that may arise once the inspector returns to the site with the warrant. This is not to suggest that OSHA will return, warrant in hand, with a vengeance.

Even though forcing OSHA to obtain a warrant can be perceived by some as an adversarial approach to cooperation, the agency and its staff are professionals who know and understand the requirements of the law. The right to request a warrant is a Constitutional right and employers who feel so compelled should exercise this right. However, depending upon the individual inspector and the nature of the impending inspection, it would be naive to think that OSHA would approach the inspection with any less scrutiny once a warrant has been obtained. The reality of the situation is that the inspection will be performed either with a warrant or without one. What is most important is how the employer handles the inspection itself.

Should a Warrant Be Requested?

As stated above, a number of different situations will determine whether an employer should request a warrant. For example, the employer may be more inclined to demand a warrant (and OSHA is more inclined to already have one) if it has a lengthy citation history, if the investigation is due to an accident involving a fatality or serious injury, or if the employer has reason to believe that it has been selected for a "wall-to-wall" inspection and is at risk of instance-by-instance penalties under OSHA's egregious policy. If a fatality has occurred, evidence of corporate and employee criminal liability may be found. Previous citations or civil violations involving the conditions leading to the fatality can be taken into account. In addition, states often improve and pursue criminal prosecutions (including reckless endangerment, manslaughter, and murder) resulting from workplace fatalities.

Therefore, when determining whether or not to request a warrant, company officials should consider the potential for criminal liability due to workplace conditions. Any employer can demand that OSHA obtain a warrant and refuse to discuss the matter (under the Fifth Amendment right against self-incrimination). While corporations do not have equal protection under the Fifth Amendment, its officers and employees do.

Challenging a Warrant

Refusing to grant entry to an OSHA compliance officer who has obtained a warrant to inspect can lead to civil and criminal contempt charges being brought by OSHA against the subject employer. When found to be in contempt of court, employers can be assessed fines and forced to pay OSHA's costs in bringing the contempt charges.

Employers may contend that their refusal to honor the warrant was based upon substantially sound reasons and in good faith, but the courts have ruled that good faith is not a defense for refusing to comply with a warrant. Remember that a judge had to make a probable cause determination before issuing the warrant. This means that the court's position is already in favor of the inspection, thereby making it extremely difficult for the employer to justify otherwise. As may be expected, employers rarely win contempt hearings. Therefore, challenging a warrant is an extremely risky strategy and should be well thought out, with advice from legal counsel, prior to any final decisions to do so.

Exceptions to Warrant Requirements

Up to this point, the discussion of warrants has focused primarily on the employer's legal rights to request them, the process that must be followed to obtain them, and the requirements associated with administering them. However, it should be noted that there are situations where OSHA can conduct an inspection without a warrant. OSHA can enter without a warrant if hazardous conditions are in plain view while inspectors are either lawfully on or off the premises. Also, OSHA may be able to perform an emergency inspection without a warrant.

Less clear but equally valid legal reasons for inspecting without a warrant are based upon the concept of *employer consent.* This is especially true for administrative inspections (as opposed to criminal searches where the standard for consent is more stringent). Failure to object to an inspection constitutes consent. Furthermore, voluntary consent by the employer to search their premises obviates any further need for OSHA to obtain a warrant, and the consent prevents the employer from later asserting that the search somehow violated their expectation of privacy. Also, less obvious but equally important is the issue of consent at multiple employer worksites. When a third party (e.g., a construction subcontractor) controls the worksite and allows the OSHA inspection to take place, virtually all employers at the site can be subjected to a warrantless inspection.

Inspection Priority: How OSHA Decides Who to Inspect

It is impossible for OSHA inspectors to visit every workplace in the United States in any given year. In fact, while it is true that OSHA is authorized to inspect every one of them, it typically inspects less than 2 percent of all workplaces per year. Moreover, although the Act covers all workplaces, the agency will not usually perform generally scheduled inspections of those sites with an average of 10 employees or less during the year.

Since their resources do not permit total inspection coverage of every worksite in the country, OSHA has established inspection priorities as follows.

Imminent Danger

Workplace situations that present an imminent danger of causing death or serious bodily harm receive the top priority. Since these conditions could lead to death or serious injury before they can be eliminated under standard enforcement actions (e.g., through programmed inspection, response to an employee compliant, etc.), OSHA is required to respond to formal complaints of imminent danger situations within 24 hours.

Catastrophes and Fatalities

After catastrophes and accidents involving a single fatality and/or the hospitalization of three or more employees, OSHA will perform an inspection. As discussed in Chapter 5 under recordkeeping requirements, employers must provide oral notification to OSHA within 8 hours following the work-related death of any employee or the in-patient hospitalization of three or more employees as a result of a work-related incident. OSHA stipulates that this requirement applies to any fatality or multiple hospitalization that may occur up to 30 days beyond the date of the incident. Hence, once the required notification has been made by the employer, a visit from OSHA can be expected.

Employee Complaints

As discussed in Chapter 7 under Employee Rights, OSHA gives its third priority for inspection to the investigation of complaints of alleged safety violations filed by employees or their designated representatives.

Figure 8-1: OSHA's order of the precedence for conducting inspection.

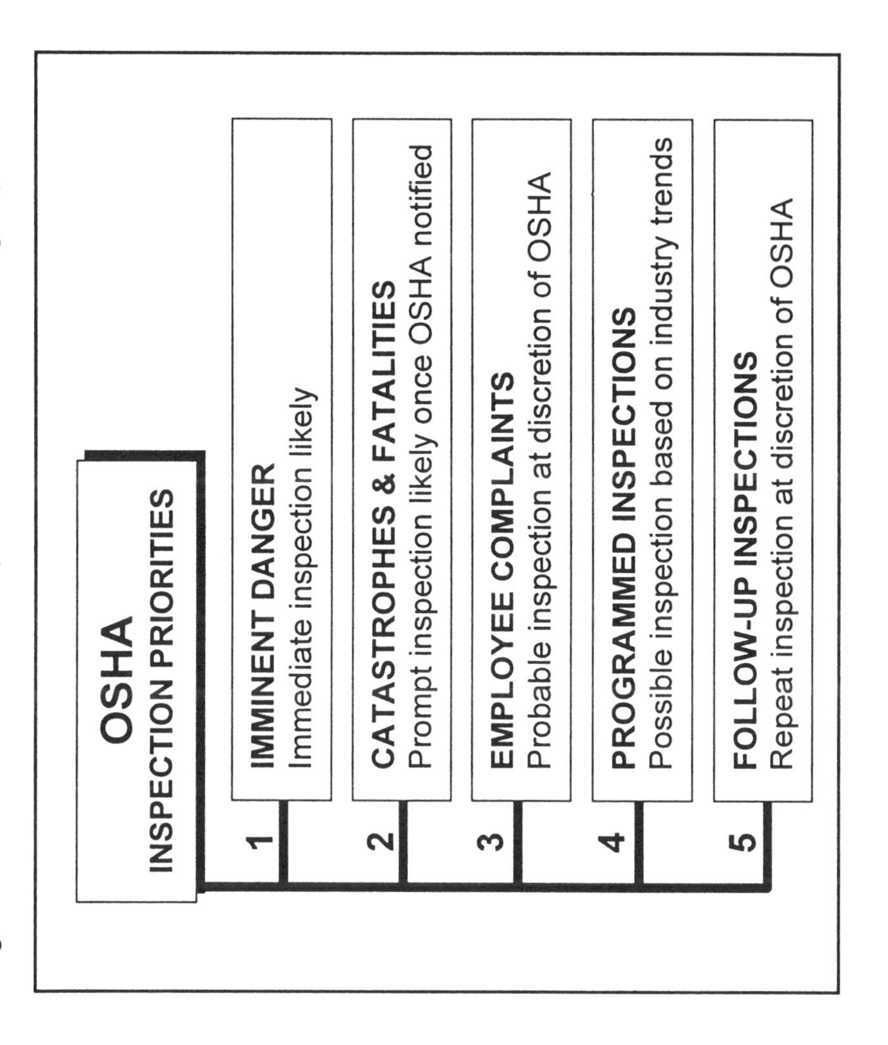

This priority is generally reserved for *formal complaints* as opposed to informal allegations. While the agency may choose to inspect the latter depending upon the nature and severity of the alleged violation(s), "nonformal" complaints will generally not provoke an inspection. In such cases, OSHA will send a letter to the employer briefly describing the alleged violation and request the employer investigate the circumstance and comply with the OSHAct (if applicable) within a specified period of time. Complainants will also receive a copy of the letter with a request to notify OSHA if the hazard is not removed. If the complainant notifies OSHA that the hazard remains, then OSHA will very likely send an inspector to investigate.

OSHA must respond to formal complaints deemed to be serious in nature (but not presenting an imminent danger) within three days. All other formal complaints must be responded to within 20 days.

Programmed Inspections

Regularly scheduled, programmed (planned) inspections are often conducted at workplaces in high-hazard industries or occupations (e.g., chemical processing, construction, etc.).

Follow-up Inspections

Last in listed priority are follow-up inspections of previously cited locations to establish whether corrective actions have been implemented to abate any cited conditions. If an employer has not corrected a violation, the compliance officer advises the employer that a Notification of Failure to Abate alleged violations will be filed. Additional daily penalties may be imposed until the violative condition no longer exists.

Special Circumstances

Although not typically listed as inspection priorities, per se, there are two other conditions which may prompt an OSHA inspection. First, OSHA may choose to conduct what it refers to as *special emphasis inspections* in industries where unusual or specific hazards may be present. For example, during 1991, the agency conducted such inspections of health care establishments and petrochemical or chemical processing plants. Hospitals and other medical institutions were targeted because of the growing hazard of HIV (AIDS) and hepatitis B. Special emphasis on the petrochemical industry was prompted by concern over OSHA's new process safety management standard and the hazards associated with non-compliance.

Second, OSHA may respond with an inspection if a particular event at a workplace (accident or incident) receives *significant media attention* to the point where OSHA becomes aware of an otherwise non-reportable occurrence. Ample television, newspaper, or radio coverage, for example, may draw the attention of the local Area Office and prompt a visit to the site, especially if the subject workplace is known to OSHA for previous safety and health concerns.

THE INSPECTION PROCESS

By law, employees have the right to request an inspection. All employees who have signed an OSHA Complaint Form (OSHA-7), or have signed a letter of complaint sent to OSHA, have the right to speak to the OSHA inspector. OSHA inspections consist of three parts: the *opening conference,* a *walk-around inspection* of the work location, and a *closing conference.* As noted in Chapter 7, the employee has the right to participate in each phase of the inspection.

The goal of an inspection is, in fact, to collect evidence to support a potential civil and/or criminal prosecution. Although the OSHA inspector is expected to be cordial, professional, and business-like throughout the entire process, the intent is far from initiating an affable relationship between OSHA and the employer. The inspector is there to investigate the employer, the employer's workplace, and the employer's safety and health programs. It is a very serious situation for the employer and for the OSHA inspector.

An inspection is not, by any definition or interpretation, a consultation. This is a critical point of fact that many employers seem to forget when faced with a visit from an OSHA inspector. All too often employers report that OSHA inspectors gave the impression that no penalties or citations would be issued if the employer corrects the conditions the inspector states are hazardous. This impression is simply not correct. According to the FOM, citations and proposed penalties are always to be issued if the inspector believes the conditions noted constitute violations of a standard or the employer's general duty obligations under the OSHAct.

General Requirements

OSHA inspections must be conducted in a reasonable manner, at reasonable times and within reasonable limits. Prior to the beginning of any inspection, the compliance officer must show proof of his or her authority (OSHA credentials, warrant, etc.) to the employer. The inspector must also give representatives of both the employer and the employees equal opportunity to participate in the walk-around inspection.

The OSHA FOM dictates the actual on-site procedures to be used by the agency during an inspection. The employer will probably not know an inspection is imminent. In fact, it is a criminal violation (see Chapter 11) of the OSHAct (and of 29 CFR 1903.6) to provide advance warning. The only exception to this rule is when OSHA specifically authorizes advance notification. Reasons for such action may be:

- The existence of imminent danger;
- The need to conduct the inspection outside normal working hours;
- In order to guarantee that employer and employee representatives will be present; and
- An Area Director believes that advance notice will produce a more efficient comprehensive inspection.

The Opening Conference

As shown in Figure 8-2, the inspection begins with the *opening conference*. This initial contact between the inspector and the employer is not simply a matter of courtesy, it is mandated by OSHA regulations. At the opening conference, the inspector presents his/her credentials to the employer or employer representative, and explains the reason for the inspection (i.e., to investigate an alleged imminent danger, a hospitalization, or programmed inspection). An employee complaint responsible for the inspection will be shown to the employer at this time. If the employee has requested anonymity, the employee's name will not be included or discussed.

Figure 8-2: The major phases of an OSHA inspection process.

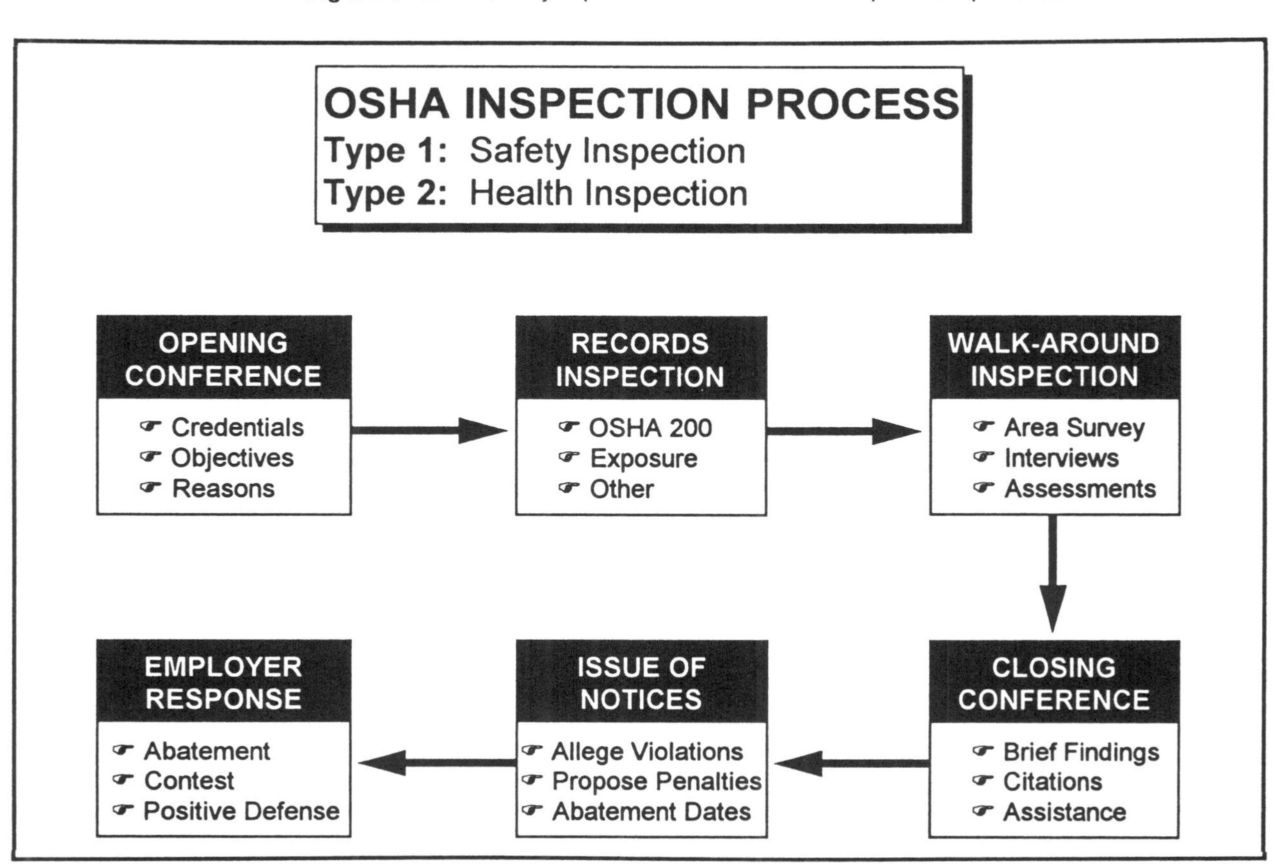

The scope of the inspection (a workplace inspection and a review of the employer's safety and health records, and possibly, private employee interviews) will be explained. The compliance officer will also provide the employer with copies of appropriate laws, standards, and regulations. During the opening conference, the inspector will establish whether any compliance exemptions or limitations apply, such as those covered by another agency's safety and health regulations.

Employee Representatives

If the employees at the site to be inspected are represented by a collective bargaining agreement, the inspector will ask that a union representative be present during the inspection. If there is no union, the employees may select a representative. The compliance officer cannot choose the employee representative, but may speak with employees during the inspection to understand the perspective of the employee population. It should be noted that the OSHAct does not specifically require the presence of an employee representative.

The OSHA Inspection Database

The compliance officer may also ask general and specific questions about the nature of the employer's activities during the opening conference. The agency maintains a database of this information to document its inspections, which is accessible to all Area Offices. Previous citations and their disposition are included. This is critical information and is especially useful to the agency when determining whether willful or repeat violations have occurred. For example, if the database shows that a company has been previously cited for violation of a given standard at one of its locations, and the same company is found in violation of the same standard at another of its locations, then a willful or even a repeat violation may be alleged.

Personal Protective Equipment

Known hazards due to operations or processes that occur in the facility should be discussed at the opening conference so that appropriate personal protective equipment (PPE) may be worn during the inspection. While the OSHAct and OSHA regulations do not specifically state that the employer must provide the inspector with the proper PPE, the employer is still responsible for the safety and health of those entering its establishment. Therefore, if the compliance officer does not have the proper equipment, the employer should provide the equipment as a matter of policy (i.e., just as it should for any vendor or other visitor entering its facility). Refusing entry because the OSHA compliance officer did not bring the proper PPE is not a justifiable defense and will hardly substantiate any good faith in compliance claims the employer may wish to make subsequent to the inspection.

Protection of Trade and Government Secrets

The opening conference is when the employer representative lists any parts of the premises where trade secrets may be seen. Objections to the taking of photographs should also be raised at this time. Employers should know that the OSHAct requires that OSHA personnel keep confidential any and all information regarding trade secrets. The OSHA FOM defines these secrets as any confidential business device or process which gives an employer an advantage over its competitors.

If applicable, the employer should advise the inspector that areas that are to be inspected are subject to federal security regulations. The OSHAct does not authorize exceptions to any other government agency's security requirements. Before inspecting such areas, the compliance officer must, therefore, have the appropriate level of security clearance. If the compliance officer does not hold the required clearance, the Area Office must provide an inspector who does.

In the past, some employers at locations where security was an issue have attempted to take advantage of a situation when an inspector does not hold the appropriate clearance to gain entry. By insisting on another inspector with the proper security credentials, the employer might gain time to correct any problems before the second inspector arrives, especially those problems mentioned by OSHA during the opening conference.

This is not to suggest that all employers who operate under government (or other) security requirements in this country are unscrupulous and dishonest. In general, the opposite is true. But, it would be naive to think that such situations have never occurred in the past. Today, however, problems with security clearances do not typically arise. Where there are business establishments that do operate under government security requirements, the OSHA Area Office is usually aware of it and, when an inspection is required, will send the properly cleared compliance officer from the onset. This avoids delays in the investigation and also helps keep our honest government contractors honest.

Employer Participation

As discussed in Chapter 6 (Employer Compliance), employers should exercise their right to participate in the inspection. Employer representatives should ask questions during the opening conference and make reasonable requests regarding how the inspection will be conducted. The inspector may even agree to a schedule suggested by the employer that will minimize disruption of company operations while ensuring the maximum value for the inspector's time.

Records Inspection

At the conclusion of the opening conference and usually prior to the actual walk-around inspection, the compliance officer will probably wish to examine the company's safety records, especially the OSHA 200 Form (Log of Injuries and Illnesses) or its equivalent, for the current year and perhaps for previous years as well. The inspector will also request the number of total hours worked for each year.

What They're Looking For

As discussed in Chapter 5 under recordkeeping, OSHA has issued regulations which specify records that must be made and maintained and which compliance officers have the right to review. They can also look at any other records which are directly related to the purpose of the inspection (e.g., employee medical records, material safety data sheets, monitoring and exposure records). Hence, the first and primary reason for reviewing the information provided on the OSHA 200 and other employer records is, of course, to verify the employer is in compliance with OSHA recordkeeping requirements (29 CFR 1904, 1910.20, and those of specific standards). Violations of recordkeeping requirements remain one of the most cited areas of non-compliance during OSHA inspections. Compliance officers routinely ask to examine employer records during a worksite inspection because such records often contain information showing the employer's knowledge of an OSHA standard and/or possible violations of that standard. These documents could then provide damaging evidence to be used against the employer.

Another reason for examining the company's injury and illness log, exposure records, and other related documents is to gain a quick understanding of any trends or problem areas that may exist in the establishment. Such information may provoke additional questions in the inspector's mind or indicate certain machines or processes that should be inspected or observed.

At manufacturing facilities, the records review allows the compliance officer to calculate a lost workday injury (LWDI) case rate for that workplace. Records review results (including the LWDI rate) are shared with the employee representative or, in the absence of such, the employer is asked to post the information where employees can refer to it.

Any applicable specific standards with recordkeeping requirements will also be examined for compliance at this time. For instance, OSHA compliance officers will almost always wish to examine the company's written Hazard Communication Program. In fact, since 1986, this has been one of the most inspected compliance requirements of any standard. Other examples may include the requirements for a written Confined Space Entry Program, a Hearing Conservation Program, or an Exposure Control Plan (as required under the Bloodborne Pathogens Standard). If Process Safety Management of Highly Hazardous Chemicals applies, the inspector will probably want to review the company's process hazard analyses.

The point is, any record or program required under the Act can be requested for review during an OSHA inspection. Employers should therefore ensure that their written programs and records are clear, organized, complete, readily available for inspection, and in compliance *before* an inspector arrives.

Are Subpoenas Necessary?

Courts have ruled that OSHA cannot compel employers to allow inspection of workplace injury and illness records they are required to keep under the OSHAct without a warrant or a subpoena because of the employer's right to privacy. But other courts decided that in producing these documents which they are required, by OSHA regulation, to make and maintain, employers have no expectation of privacy. OSHA has taken the position that it is a reasonable assumption that an employer would have less of a privacy interest in a document it is required to maintain by regulation or statute than in a document it produces and maintains on its own. Regardless, employers should be aware that compliance officers do have the power to subpoena employer records, if necessary.

The Workplace Walk-Around

After the records review, the *walk-around* inspection begins. "Walk-around" is OSHA's term for walking through the workplace and obtaining a mental "snap-shot" view of the employees' activities and working conditions. In some cases, depending upon the individual inspector and the reasons for the inspection, this will conclude the inspection. But, it is more common for the inspector to ask questions of the employer and of employees, and make notes regarding areas or items to explore further during a subsequent inspection.

General or programmed *safety inspections* are walk-around inspections. *Health inspections* include a walk-around and possibly some type of health monitoring (e.g., to determine airborne contaminant concentrations, noise levels, radiation exposure, etc.) if excessive levels are suspected.

During a more in-depth inspection, the compliance officer may take photographs, question or interview employees, and take a variety of measurements. These actions are dependent upon whether a safety or health inspection is being conducted, the information the OSHA inspector has obtained up to that point, the employer's standard practices, an employee complaint, the inspector's distance from home, and even the time of day, along with many other less quantifiable factors the inspector evaluates on a case-by-case basis, either consciously or unconsciously.

Safety and Health Inspections

As stated previously, there are essentially two types or categories of OSHA inspections. OSHA developed this approach to properly focus its resources on the specific problems known or suspected (alleged) to exist at a given workplace. While there are certainly some OSHA compliance officers who are trained in both areas of expertise, it is more common to encounter either a safety specialist or a health specialist (industrial hygienist, usually) during an inspection.

The Safety Inspection. Safety inspections begin with gathering information on hazards, observing employees at work, and, where appropriate, interviewing employees. The next step in this process is to evaluate the employer's safety program with regard to its level of compliance, its effectiveness in the workplace, degree of employee participation, completeness and adequacy, and other related concerns. All the facts relating to any apparent violations discovered during the inspection, along with the inspector's observations and opinions, will be recorded and brought to the attention of the employer.

The Health Inspection. Overall, the basic health inspection procedures are the same as those described for the safety inspection. However, they are usually more detailed and are frequently conducted by industrial hygienists. At the beginning of the health inspection, the compliance officer must give the employer representative a letter stating why the workplace was selected for the health inspection. Usually such inspections are subsequent to a formal employee complaint. However, the employer may have been selected for inspection because of its standard industry code (SIC) classification, because it is in a high hazard industry, or it falls under an OSHA special emphasis program. Whatever the reason, employers have the right to know why they are being subjected to health inspections.

The compliance officer then reviews all relevant workplace records, looking for insights into any frequently documented symptoms or trends (which may indicate workplace exposures to toxic substance), frequency of illness or injury (which may indicate the severity of the problem), specific conditions and their frequency (cases of dermatitis, for example), use of personal protective equipment, results of any previous exposure monitoring or tests (audiometric tests, air sampling, radiation monitoring, ventilation tests, illumination, etc.), process flow charts of hazardous operations and tasks, and a listing of hazardous materials present in the facility. The walk-around may be supplemented with air sampling and evaluation of the employer's safety and health program. The effectiveness of engineering controls being used will be determined. Samples will be used to ascertain the need for more extensive sampling.

Conduct During the Inspection

It is extremely important that the employer representative remain with the compliance officer at all times during the inspection (except, of course, when the inspector is conducting private interviews with employees).

Ask Questions

The employer should ask questions of the inspector and question any methods or techniques being used that the employer does not understand. The employer may also wish to ask for the inspector's opinion about the safety and health practices and conditions observed during the walk-around. While most inspectors may refrain from rendering such an opinion, it causes no harm to ask.

Any responses could assist the company in planning improvements and, more importantly, might be useful in later conferences or trial. Such exchanges may also give insight into the inspector's familiarity with the employers' industry. Knowledge of an inspector's ignorance in a given subject area can be useful especially in challenging a citation the employer believes to be unjustified or unfounded. Employers should also make their own notes and observations throughout the inspection, especially when the compliance officer seems particularly interested in a given process, procedure, task, operation, piece of equipment, or other subject area.

Of course, the inspector will also be asking the employer questions throughout the inspection and taking notes. If there are ever any proprietary, legal, or other concerns at any time during the inspection, the employer's representative has the right to request a "time-out" in the inspection to confer with the company attorney or other appropriate representative of management. This is to protect against self incrimination.

Watch What You Say

The employer's statements can be used during a contest or other hearing. Not all employers are totally versed on all laws related to such matters and OSHA understands this.

Interrupting an inspection process for a short period of time so that legal counsel or management guidance can be obtained is a totally acceptable and appropriate request for an employer to make. If, however, the employer neglects to do this and freely provides comments, statements, observations, or any other type of opinion to the compliance officer during the inspection, then employers may hear their own words being used to substantiate claims of noncompliance during some subsequent court proceeding. In other words, if the employer's representative carelessly decides to voice an opinion or make a statement that seems to affirm claims of noncompliance, they should not expect the compliance officer to warn them against making such statements. The compliance officer is more likely to respond by listening very carefully to every word and noting any comments made by the employer. The most damaging evidence presented against a company during trial often consists of statements made by management representatives to the inspector.

Photographs, Documents, and Other Evidence

The employer must always remember that the primary objective of any OSHA inspection is to gather *evidence*. The compliance officer is there for the specific purpose of obtaining evidence to support the fact that violative conditions exist. This evidence can consist of company documents, including the results of self audits and reports prepared by consultants or insurance companies, employer and employee statements, photographs, video and audio tapes, and sampling data. Articles that address a particular hazard appearing in trade association publications, safety meeting minutes, employee complaints, internal correspondence between safety personnel and corporate management, and any other related documents can all be obtained by subpoena if the inspector believes they are in existence but are not made available upon request.

When photographs are taken by the inspector, he or she usually has one of two reasons to do so. *General,* or non-specific, photographs will be used to show an overall view of the area under inspection. *Specific* photographs will be used to document individual violations or conditions. Also, the trend in recent years in OSHA inspections is the use of videotapes to record violations encountered during an inspection. This is especially true at construction and other multi-employer worksites. Whatever the method used by the inspector to gather photographic evidence, the employer should be prepared to do the same. That is, the employer should take the same photographs and make the same videotapes at the same moment in time and under the same conditions (lightning, weather, employee shift load, equipment operation, time of day, perspective and angle, depth of field, and so on).

Since OSHA will generally not provide the employer with copies of any photographs taken during the inspection, taking their own is the best way to document the same information as the inspector. Also, having their own copies may help counter any unfounded allegations that may be made by the inspector at some point in the future.

The Employee Interview

As stated previously in this text, OSHA compliance officers conducting OSHA inspections are authorized to question employees privately. This authorization is obtained through the OSHAct and two separate OSHA regulations at 29 CFR 1903.3(a) and 1903.7(b). Employers should known that OSHA cannot simply conduct these interviews at any given time and in any way they see fit. These interviews must observe reasonable limits and be conducted in a reasonable manner. They should be as brief as possible so as to keep any disruption to employers' operations and employees' duties to an absolute minimum.

According to the OSHA FOM, the purpose of the interview is to assist the compliance officer in obtaining information that may be necessary or useful in carrying out the inspection effectively. Also, if it should become necessary, both the Act and the FOM authorize OSHA to issue administrative subpoenas for the purpose of compelling employee interviews.

In making sense of this area of compliance, it is important that the employer understand that statements made by any of its own employees can be used to support OSHA's case during trial. Such statements will be considered statements of fact, as opposed to hearsay, if they were made during the existence of an employer-employee relationship.

The criticality of of statements by these employees, as well as by employers, cannot be overestimated. While OSHA inspectors do not normally take signed statements (although there is nothing to prevent them from attempting to do so), they usually record what is said to them in their field notes. Even though the notes are not usually verbatim, it is still a fact that more violations arise from what is said during the inspection, either by the employer (including documents voluntarily provided by the employer) or by employees, than from what the inspector actually sees on-site. For this reason alone, employers are urged to plan ahead, preparing policies and procedures for dealing with such issues well in advance of any inspection.

The Closing Conference

When the walk-around inspection has concluded, the inspector confers with the employer and employee representatives. While closing conferences are important, like opening conferences, the failure to hold one will not result in the vacating of a citation unless there has been prejudice to the employer. In other words, if the compliance officer neglects to hold the closing conference, it is not grounds for dismissing any citation unless the employer can show that the lack of the conference resulted in unfair, partial, and/or prejudicial consequences to the employer. This is true, by the way, of the opening conference as well.

During the closing conference, the inspector will usually present observations and comments on workplace conditions and practice. This may indicate whether or not there is a possibility that citations will be issued. There may also be additional questioning of the employer to support the inspector's later actions. Therefore, employers are urged to be extremely cautious at this point that they do not make statements that are tantamount to admissions of guilt or noncompliance. This is not suggesting that OSHA will purposely attempt to trick or entrap the employer. It is only meant as a note of caution. Just because it is called a "closing conference" does not mean the inspector's evidence-gathering objective has ended. The closing conference is not the time for employers to let out a sigh relief and lower their guard.

The compliance officer should also explain the next steps that might be taken in this process. For instance, OSHA should inform the employer that if a citation is issued, follow-up inspections may be used to confirm that the citation had been posted and that workers are not being exposed to the cited hazards during the abatement period.

In some instances, especially in health inspections, the closing conference may not occur until several weeks after the conclusion of the on-site inspection. For example, air samples taken while monitoring for toxic substances are sent to the OSHA laboratory in Salt Lake City, Utah or Milwaukee, Wisconsin for analysis. The inspector must receive the results before concluding that a violation has occurred. Other delays are due to OSHA's internal operating procedures, which require higher-level approval before a local Area Office can issue "willful" citations or when there are special conditions associated with other types of citations.

It should go without saying that employers should not argue with OSHA compliance officers or attempt to negotiate the company out of a citation. Such foolishness could result in an unnecessary confrontation or might disclose additional damaging evidence. Such arguments are less risky and are more likely to achieve success if made at a later date.

INSPECTIONS AT CONSTRUCTION SITES

Because multiple employers are often involved, construction inspections are often quite complex affairs.

Consider these important issues:

- Third parties can consent to searches of jointly occupied property, if the third party has what is known as *common authority* over the location and, therefore, can give permission to the compliance officer to inspect the construction site. A general contractor, for example, is considered to have such authority. When permission is given by the common authority, subcontractor right to privacy is virtually eliminated. Also, the common authority is supposed to call all subcontractors together at the opening conference to inform them of the inspection. While a subcontractor who has not been so notified will generally not be subject to inspection, it does not mean total exemption from citation if violations are discovered.

- Even without permission to enter the construction site, the *plain view doctrine* allows citation of employers for violative conditions the compliance officer can observe from a public place. This means that the OSHA inspector can watch construction activity from outside the boundaries of the worksite and note any violations that occur. Further, the inspector can take photographs or videotape (using zoom features if necessary) for use as evidence against the employer(s).

- The courts have ruled that employers responsible for a hazard may be found in violation of the general duty clause if their own employees *or* other employers' employees on the construction site are exposed. In other words, a subcontractor will be liable for violative conditions under the GDC if employees of other subcontractors are exposed, even though none of its own employees are exposed to the hazard. OSHA need only prove there is access to the *zone of danger,* not actual exposure, to substantiate that a violative condition occurred. Pictures and videotape are quite effective in this process.

- Every employer whose employees were exposed to a hazard can be cited, regardless of whether another employer was contractually responsible for controlling it. Subcontractors can avoid such potential by:

    a) Taking appropriate measures to protect their employees against the hazards, or;
    b) Proving that they did not have the expertise to recognize that hazardous conditions existed.

  For example, when a hazard is in fact recognized as such, subcontractors may avoid citation by showing (through documented letters, meeting minutes, or some other means) that they took their concerns to the general contractor for resolution. Subcontractors should ask general contractors and responsible subcontractors who created the hazards (if known) to abate them. But this is not considered sufficient if alternative means of protection are available. It must always be remembered that OSHA's primary interest lies in the protection of employees. If nothing is being done to ensure this objective during the time when subcontractors and the general contractor are negotiating a fix or blaming each other for creating a problem, then OSHA will still look to all responsible parties involved for citation. Protecting employees in the interim can be as easy as warning them of the hazard and instructing them to avoid it until it is corrected.

- The issue of general contractor liability often arises at multi-employer worksites. Specifically, should a general contractor be held liable for violations created by its subcontractors? The answer is, it depends. The OSHRC has held that the general contractor is responsible for subcontractors' violations if those violations were such that ordinary supervision over the subcontractor would have detected, prevented, and abated the conditions. Specifically, the courts have held that general contractors normally have the responsibility and the means to assure that other contractors fulfill their obligations with respect to employee safety. Hence, the duty of a general contractor is not limited to the protection of its own employee from safety hazards. It extends to the protection of all the employees engaged at the worksite.

When establishing whether a general contractor is responsible for its subcontractors' safety violations, three factors should be considered:

- The degree of general contractors' supervisory authority over the subcontractors;
- The severity of the violation; and
- The precautionary measures taken to protect potentially exposed employees.

Chapter 9

VIOLATIONS, CITATIONS, AND PENALTIES

OVERVIEW

Issuing *penalties* subsequent to the finding of *violations* and issuance of *citations* is and has been the primary method used by OSHA to enforce the provisions of the OSHAct. Citations are issued and monetary penalties imposed under Sections 9 and 17 of the OSHAct. During an inspection, the OSHA compliance officer is especially interested in determining the employer's level and adequacy of compliance with applicable OSHA standards and/or the general duty requirement. The whole purpose of the inspection is to gather enough evidence to prove whether or not a violation(s) of OSHA standards exists. When such is the case, OSHA issues a citation indicating by exact reference which of its standards are alleged to be in violation. The issuing of citations may or may not be accompanied by proposed penalties which the employer must accept or appeal.

This chapter will discuss OSHA citations, the agency's method of determining penalties, and how the issue of *employer knowledge* (or lack thereof) plays a significant role in the entire process. The *categories of violations* include repeat, willful, serious, other than serious, failure to abate, and de minimis. The *types of penalties* and penalty reduction factors will also be discussed.

INTRODUCTION

Throughout this text, there have been numerous references to and discussion about the fact that OSHA finds employers in violation of regulations or standards and issues citations which may contain proposed penalties pursuant to their authority under the OSHAct. Issuing penalties subsequent to the finding of violation and issuance of citation is and has been the primary method the agency uses to enforce the provisions of the Act.

During an inspection (see Chapter 8), the OSHA compliance officer is especially interested in determining the employer's level and adequacy of compliance with applicable OSHA standards and/or the general duty requirement. An inspection is conducted to compile evidence establishing whether or not OSHA standards have been violated. OSHA issues a citation stating which standards have allegedly been violated and may also impose penalties, which the employer must accept or appeal.

EMPLOYER KNOWLEDGE

Section 17(k) of the OSHAct specifically states that a serious violation can only be alleged if the employer knew, or with the exercise of reasonable diligence, could have known of the presence of the violative condition. Although the stated requirement is for serious violations, the OSHRC has ruled and the courts have held that knowledge of the violation is an essential requirement of proof for other than serious violations as well. In fact, employer knowledge of a violation is required for OSHA to substantiate its citation allegations and for fines to be imposed in almost all cases. In other words, the Act does not impose strict liability.

This *knowledge* is not easily defined. There are not only degrees of knowledge, there are also a variety of ways by which it can be gained. For example, for a willful violation, the cited employer must have actual knowledge of the existence of the hazard *and* that either the condition is in violation of a safety standard or that the employer simply disregarded employee safety.

Another somewhat ambiguous knowledge requirement is applied to both serious and non-serious violations. It states that if the employer knew or reasonably could have known of the existence of the violative condition, then the knowledge requirement is fulfilled even though the employer may not have actually known that an OSHA standard had been violated.

How Employers Gain Knowledge

According to OSHA, employers acquire knowledge through a wide variety of ways beyond the obvious direct knowledge which is typically demonstrated by showing that an employer was advised that a hazard was present. Employer knowledge is most frequently based upon *respondeat superior,* a rule of law under which employers are responsible for employees' actions and/or the *principles of agency* that govern employer/employment relationships. The principles of agency that apply here hold that a principal (the employer) is responsible for the acts of any of its agents (employees) that are within the scope of the *agency relationship* (employment). For example, if OSHA determines that a supervisor personally knew of a violative condition but never passed the information on to upper management, OSHA would still find adequate evidence of employer knowledge under the principles of agency and/or respondeat superior.

The "Reasonable Diligence" Factor and Employer Knowledge

The OSHAct specifically states that a serious violation can not be alleged as having occurred if the employer did not and could not, with the exercise of *reasonable diligence,* know of the violation. Essentially, this language means that an employer is liable for conditions or practices which should reasonably have been known of and/or for taking preventive actions. In other words, OSHA believes employers must monitor their workplace safety and become knowledgeable of all hazards that may be present there.

What is considered "reasonable" diligence? While there are no clear answers in the case law, the rulings seem to indicate that those employers who act responsibly and make workplace safety and health a visible priority in business operations are considered to be exercising reasonable diligence over their safety programs. If the alleged violative condition would have been discovered under such a program, then it is assumed that the exercise of reasonable diligence would have prevented the violation in the first place. Clearly, the application of this logic may not always be sufficient to prove reasonable diligence. Therefore, the courts will always examine each case of alleged safety violation in consideration of all pertinent facts related to the determination of reasonable diligence. All this essentially means is that employers must do everything reasonably possible to protect their employees. If OSHA can determine that the employer was not diligent in providing a safe and health workplace, then the agency will most likely pursue a citation.

Using Lack of Knowledge as a Defense

Obviously, with all this emphasis placed on the employer knowledge requirement, it might be assumed that employers can claim lack of knowledge as a defense when contesting a citation. Such defenses are divided into two categories. The most common type based upon *employee misconduct.* The others are based on misinformation or the employer's *inability to determine* the hazard's existence.

Employee misconduct is considered a "good-faith' defense (i.e., the employer must have behaved correctly). To establish an affirmative defense based upon unpreventable employee misconduct, employers must prove that the actions leading to noncompliance with the standard were unknown to the employer *and* were contrary both to employers' specific instructions and to uniformly enforced company work rules equivalent to the standard allegedly violated. In most cases, the employer must also demonstrate that the employees were adequately trained and supervised with respect to the specific workplace safety hazards that are associated with their occupation.

In addition to the above, several factors are typically considered when attempting to apply the employee misconduct defense, including the employers' *compliance history.* The courts will look at whether the employer's past violations show insufficient enforcement of safety standards. Conversely, a history of infrequent citations often indicates a company's stringent observance of standards. Therefore, it would be prudent policy to ensure that appropriate and fairly administered disciplinary actions are consistently imposed on employees who fail to comply.

The Literal Ignorance Defense

Lack of employer knowledge can also be a defense when the employer's *literal ignorance* of a hazard prevents the employer from protecting workers against it. Unforeseeable accidents or incorrect information presented to the employer regarding the existence of the hazard are examples of literal ignorance.

For example, employers often rely upon the information presented in a manufacturer's material safety data sheet (MSDS) to be an accurate assessment of health hazard risk. They will typically take the necessary precautions to prevent employee exposure to those stated hazards per the manufacturer's recommendations. These actions are considered adequate and, in fact, are quite common practice in assessing and preventing employee exposures. If, however, the information provided to the employer proves incorrect or misleading, thus concealing the true hazardous nature of the chemical product, then the literal ignorance defense may prove successful.

Employee Leasing Considerations

A word of caution is warranted when discussing the lack of knowledge defense with respect to the employer's absolute duty to care for employees. This duty is not delegated when one employer loans or leases an employee to another employer. The first employer will be held liable if it does not determine what the third party employer's tasks entail. This is because the leased employee typically remains the leasing employer's responsibility (paying wages, etc.). So, the leasing employer is responsible for ensuring their safety.

CATEGORIES OF VIOLATIONS

There are a number of violation types or categories which OSHA can cite depending upon the situation the agency encounters. These are as follows.

De Minimis Violations

In summary, de minimis violations are defined in the OSHAct as violations which have no direct or immediate relationship to safety or health. The courts have ruled that an OSHA standard violation is classified as de minimis rather than non-serious if there is no clear relationship between non-compliance and employee safety or health. Such citations do not require abatement or payment of penalties. De minimis violations have also been alleged where the possible injury involved is of a minor nature not requiring abatement.

De minimis violations may include procedural noncompliance. These are situations when the employer may not be in strict compliance with a particular standard but has implemented alternate means of protecting the employee. In other words, the employer may not be in compliance with the strict letter of the law, but is still ensuring that the intention of the law is met. Hence, if a non-compliant situation exists, but there is no harmful result to employees, the violation is considered de minimis.

Technically speaking, a violation which presents no threat of harm should be classified as de minimis or the citation should be vacated. There is really no safety or health benefit for situations of this kind, but OSHA persists in issuing these citations and the OSHRC persists in affirming them.

Other Than Serious Violations

Other than serious (OTS) or "non-serious" violations exist when the possible results could be minor illness or injury (i.e., illnesses or injuries which do not meet the serious physical harm criteria). If proof is not available (i.e. possibility of death or serious physical harm is lacking), OSHA will usually issue an OTS violation.

Serious Violations

The OSHAct states that a serious violation exists when there is a substantial probability that death or serious physical harm could result from conditions, practices, means, methods, operations, or processes which exist, or are in use, in the workplace. To allege a violation is serious, OSHA need only prove that an accident is possible and that serious injury or death would be its likely outcome. Chapter 4 of the OSHA FOM establishes the agency's criteria for serious physical harm. Examples include:

- Amputations (loss of all or part of an appendage, including the loss of bone);
- Burns or scalding (including electrical and chemical burns);

- A cancer;
- Concussions;
- Crushing (internal, even though skin surface may be intact);
- Cuts, lacerations, or punctures causing significant bleeding or requiring sutures;
- Fractures (simple or compound);
- Hearing loss;
- Lung diseases (such as asbestosis and silicosis); and
- Poisoning (due to inhalation, ingestion, or absorption through the skin of a toxic substance adversely affecting a bodily system).

In addition, the FOM further defines serious harm as either, 1) the permanent or temporary impairment of the body due to part of the body being rendered functionally useless or substantially reduced in efficiency; or, 2) illness that could shorten life or significantly reduce physical or mental efficiency. The potential for harm determines whether a violation is serious.

Willful Violations

Just as with repeat violations, no clear definition of willful violations is supplied in the OSHAct. However, with respect to the employer knowledge requirement discussed earlier in this chapter, a violation becomes a willful violation when the violation results from intentional disregard or utter indifference to the workplace safety codes. Specifically, it is the employer's intent to disregard a regulation, rather than their rationalization or motive, which is relevant in determining willfulness. Regardless of any facts which may support an employer's non-compliant actions as necessary (to prevent economic loss, for example), if the employer knew of the regulation and decided not to comply, it is a willful violation. Therefore, it is not the fact that the employer has recognized the hazard, but that there is *employer knowledge* of a standard which is being violated, that establishes the willful nature of the violation. The employer knows a standard exists and willingly chooses to ignore it, violating the law. Therefore, a willful violation can result from *any* violation of an OSHAct provision or OSHA standard.

There are three basic elements of a willful violation that OSHA must prove to sustain any allegation of employer willful misconduct. All are based upon the concept of employer knowledge. First, OSHA must show that the employer was aware that a standard was being violated. Second, evidence that the employer knew the regulation in question required the employer to take actions to eliminate the condition or practice must be provided. Third, OSHA must show that the employer decided to ignore this responsibility despite full knowledge of its existence. It is often difficult for OSHA to conclusively establish the employer's state of mind regarding the infraction. As discussed previously, one way OSHA can demonstrate employer knowledge of a violated duty is to demonstrate that the employer had violated that regulation previously, and that citations for the same condition had been issued previously. It should be noted that it is not considered a willful violation if the employer, in good faith, misinterprets the law or has a genuine dispute over the meaning of the law.

Failure to Abate Violation

If the actions required to abate a cited hazardous condition are either never started or only partially completed within the time period established, OSHA issues a failure to abate notice to the employer. The notice may propose penalties for each day the alleged violation continues to exist. The OSHAct authorizes daily penalties of up to $7000.00 for failure to abate situations.

If a follow-up inspection finds the identical violative condition that was the subject of the original citation and OSHRC final order, the failure to abate is proven. The violation also applies if the follow-up inspection finds that the condition was only partially abated. This will prevent any subsequent failure to abate penalty against the employer, particularly if the employer can demonstrate that it was proceeding in good faith to accomplish abatement and simply needed more time to finish. In such a case, the employer might counter plead with a late-filed petition to modify the abatement period.

To substantiate the claim of failure to abate, OSHA must also show that the hazard existed continuously and was never corrected from the initial inspection through the follow-up inspection. A failure to abate violation would not apply if the hazard was abated at any time during this period, but re-arose for whatever reason. On that note, prevention of worker exposure to the cited hazard would be considered effective abatement even if the hazard continued to exist. An employer may be able to dispute a failure to abate citation by proving the original citation to be invalid, even if it was not contested. For instance, the original citation may have included a condition which did not violate the OSHAct or the original citation may not have clearly stated the type of abatement imposed.

Penalties. As shown in Table 9-1, willful violations carry significant consequences, with minimum penalties of $5,000.00 rising to $70,000.00 for each civil willful violation.

Time Period Considerations. According to the OSHAct, the time period for an employer to abate a violation begins when a final OSHRC order becomes effective. Therefore, by filing a notice of contest in good faith within fifteen days from the issuance of the citation, the employer will effectively prevent the possibility of a failure to abate charge during the original citation's review. A failure to abate citation can be issued if no contest was filed during this fifteen day period or if the circumstances of the original citation required immediate abatement and no such abatement actions were taken.

Repeat Violations

Although the OSHAct prescribes penalties for repeated violations (see Table 9-1), it does not establish the meaning of the violation itself. Case law, however, indicates that a violation is considered "repeated" if the same standard has been violated more than once by the same employer and there is a *substantial similarity of violative elements* between current and prior violations. This means that the previous citation must have been for an offense that is essentially the same as the repeat violation. For example, if an employer is cited for failure of its employees to wear appropriate foot protection, it is not considered a repeat if the same employer is cited for failure of its employees to wear hard hats, even though the same standard is in violation (29 CFR 1910.23). Because the prior and current violations are not substantially similar, the second is not considered repeat. Also, the previous citation must have been a *final order* by the OSHRC (i.e., case closed on ruling of the OSHRC) in order to allege a repeat of that violation at some time in the future.

Table 9-1: Violations and penalties under the OSHAct.

CATEGORY	PENALTY
Repeat Violation	At least $5000.00 and up to $70,000 **per** violation
Willful Violation (non-death)	At least $5000.00 and up to $70,000 **per** violation
Willful Violation (causing death)	Up to $10,000.00 and/or six months imprisonment **for a first conviction;** $20,000.00 and/or one year imprisonment for each subsequent violation*
Serious Violation	Up to $7000.00 **per** violation
Other Than Serious Violation	Up to $7000.00 **per** violation
Failure to Correct Violation	Up to $7000.00 per day the violative condition continues
Giving Advance Notice of an Inspection	Up to $1000.00 and/or six months imprisonment
Providing False Statements on Documents Required to be Filed Under the Act	Up to $10,000.00 and/or imprisonment for up to six months

* *Willful violation resulting in death is a criminal offense. Corporations can be fined up to $500,000.00 per violation for willful noncompliance (see Chapter 11 on Criminal Prosecutions for more detail).*

FOM Guidelines for Repeat Violations. The OSHA FOM provides guidance to governing the issuance of repeat violation citations. First and foremost, OSHA will look only to the previous three years to determine if the alleged violation is a repeat of some previous concern. It is noted here that these are only guidelines and not law. OSHA can choose to ignore any of them if they believe it serves the purpose. For example, citations issued beyond the three year mark are often used to determine whether the current violation is willful.

The FOM guidelines also establish geographical limits on the determination of prior violations. Theoretically, this was so employers who operate in more than one state or region could not be held for a repeat of a violation that occurred in some other location. However, OSHA no longer takes geographical limits into consideration in issuing citations. Employer worksites are now classified into three types. *Fixed establishments* are offices, factories, restaurants, retail stores, and the like. Consideration for repeat violations in such cases is limited to the fixed establishment being cited. If, at some point in the future, OSHA discovers the same violation exists at a company-owned/operated fixed establishment elsewhere, then the company may be fined for repeat violations. Citations related to *non-fixed establishments*, which include construction sites and oil and gas drilling sites, will be compared to all sites in the OSHA Area Office's jurisdiction. Citations for violations at *long shoring establishments* relate to all long shoring activities of a single stevedore at the port in question.

In general, a violation by the same employer of the same standard is to be regarded as a repeat violation and that one prior violation is enough to support a finding of repeated violation.

Egregious Misconduct

In addition to the statutory violations described in the OSHAct, a new type of violation for *egregious misconduct* has been derived from the willful violation category. An egregious violation occurs when OSHA finds non compliance with the same standard in more than one instance or at more than one location but under the control of the same employer. Egregious violations are cited on an instance-by-instance or violation-by-violation basis.

CITATIONS AND PENALTIES

Citations and penalties are determined by the OSHA Area Director after receipt of the compliance officer's inspection report.

Citations

The compliance officer normally advises employers of apparent violations that were discovered in or at the inspected location at the conclusion of the closing conference or at sometime after that (delays while waiting for sample results or photograph development are not uncommon). The inspector will then prepare one or more citations describing the alleged violations and proposed penalties, in consultation with appropriate OSHA supervision.

Some types of citations require higher level approval. For example, OSHA Regional Office approval may be needed for general duty clause violations. The National Office and the Office of the Solicitor usually must approve egregious violations.

The OSHAct requires that citations be delivered to employers promptly. Notices of citations are usually mailed to an employer by certified mail, with a copy sent to the employee union or other designated representative. OSHA uses a return receipt to establish the beginning of the contest period. The citation informs the employer and the employee of violations of regulations and standards, and states a time period to correct (abate) such violations.

Under the provisions of the OSHAct, OSHA is authorized to issue citations if they *believe* a violation of an OSHA standard or the general duty clause has occurred. This means that the OSHA compliance officer does not have to *prove* that a violation occurred; a reason to believe one existed is enough to justify a cite.

The Particularity Requirement. According to the OSHAct, citations must be in writing and include very specific and particular detail on the nature of the alleged violation together with reference to the provision of the OSHAct or standard that has allegedly been violated (this is required for all but de minimis violations). This *particularity requirement* ensures that employers have enough time to decide whether or not to contest the citation, and that the violation is described in sufficient detail for the employer to actually identify the alleged violation. Lack of particularity has been upheld as defense to a citation, but it must be raised during the early stages of a proceeding (e.g., during the pleadings), or it will be considered waived by the courts.

Notices. When a particular violation is alleged to exist that is considered extremely minor in nature and the results of continued non-compliance would be transparent (i.e. have no effect) with regard to worker safety and health, the agency can issue a *notice* rather than a *citation.*

Forms. Almost all citations are issued on a computer-generated form. This form uses *standard alleged violation elements (SAVEs)* and *alleged violation descriptions (AVDs).* SAVEs and AVDs standardize and expedite the issuance of citations.

Posting Requirements. With any citation, the law requires the employer to post a copy of it at or near the place where the violation is alleged to have occurred. Posting must be for three days, or until the violation is corrected, whichever is longer. Where OSHA allows an employer longer periods to abate (three months, for example), a union representative or safety and health committee member may request a copy of the employer's abatement plan and a copy of the three-month progress report which is required by OSHA.

Statute of Limitations. It is also extremely important that employers be aware of the Section 9(c) provisions of the OSHAct. Specifically, there is a six month statute of limitations for issuing of a citation from the date of the alleged violation's occurrence. This means that OSHA must issue the citation within six months of first discovering the violation. Also, if an employee has filed a complaint alleging a violation, OSHA is not obligated to respond if the employee claims the violation occurred six or more months prior.

Penalties

Under Section 17 authority, OSHA can propose penalties for violations of the Act. Table 9-1 summarizes the penalties that are linked to the various categories of violation under the OSHAct.

Table 9-2: The catagories of OSHA violations.

SEVERITY/GRAVITY DETERMINATION		
CATEGORY	CHARACTERIZATION	POINTS
1	Non-Serious	0
2	Serious, but minor, easily treated consequences	1-3
3	Serious, with treatable, non-permanent consequences	4-7
4	Serious, with permanent consequences or death	8-10

ACCIDENT PROBABILITY DETERMINATION		
FACTOR	CLASSIFICATION	POINTS
Exposed Employee	Per Worker, up to 10	1 (per employee)
Exposure Frequency	Up to once a week	1-3
	Up to Daily	4-7
	Continuous Daily	8-10
Employee Proximity	Fringe of Danger Zone	1-3
	Near Danger Zone	4-7
	At Point of Danger	8-10
Working Conditions (pace, noise, etc.)	Low Stress, Good	1-3
	Medium Stress, Fair	4-7
	High Stress, Poor	8-10

HEALTH RISK DETERMINATION		
FACTOR	CLASSIFICATION	POINTS
Exposed Employees	Per Worker, up to 10	1 (per employee)
Exposure Duration	1-8 hours/week	1-3
	More than 8 hours but not continuous	4-7
	Continuous Daily	8-10
Personal Protective Equipment Use	By All Exposed, With A Good Program	1-3
	By Some Exposed, Slight Program Deficiencies	4-7
	By No One	8-10
Medical Surveillance	Effective	1-3
	Partially Effective	4-7
	No Surveillance, or Ineffective	8-10

Serious violations require imposition of penalties. A formula for calculating penalties is used. It applies a numerical value to the categories of violation. As shown in Table 9-2, four categories of violations are assigned a corresponding number of points. Penalties are assessed based upon the following factors.

The Gravity of the Violation. The OSHA FOM establishes formal guidelines for penalty calculation. These guidelines, as summarized in Table 9-2, are based on formulas which incorporate the number of workers exposed, frequency of exposure, potential risk of adverse effects, and their severity/gravity. The results of this initial assessment will yield a *probability* and *severity* value for the particular citation. Once these are determined, the compliance officer will determine the probability/severity (P/S) quotient (the average of the two values). This process, often referred to as the *gravity-based* proposed fine, then uses a penalty table (see Table 9-3) which factors in percentage reductions up to 80 percent, depending upon the evaluation of the next three factors.

The Size of the Employer's Business. As shown in Table 9-3, the proposed fine can be reduced by a percentage amount that directly responds to the number of employees at the cited establishment. However, those employers with greater than 100 employees can expect no reduction based upon this factor.

The Good Faith of the Employer. For those employers who can demonstrate a good faith in compliance effort in their facilities, OSHA will allow up to a 30 percent reduction of a proposed penalty. By simply having a well documented and well implemented occupational safety and health program in place, an employer can substantially reduce a proposed OSHA penalty.

Prior History of Violations. If OSHA finds very little historical information with respect to past violations and citations, they are at liberty to reduce a proposed penalty by a maximum of 10 percent in this category (Figure 9-4).

Table 9-3: OSHA's Penalty Table.

PENALTY REDUCTION FACTORS		
FACTOR	**REQUIREMENTS**	**PERCENTAGE**
Number of Employees	1-10	40
	11-25	30
	26-60	20
	61-100	10
Good Faith in Compliance	Effective Safety Program, Few Violations, Injuries	30
	Relatively Effective Program, No Willful or Repeat Violations, Few Serious Illnesses or Injuries	20
	Some Safety Program, Abatement With Just Acceptable Promptness, No Willful, Repeated Serious, or Serious and High Gravity Violations	10
	Willful and Repeat Violations, Easily Prevented Serious Illness	0
Employer History	General Factors, Past Inspections, Prior Citations, Categories, Etc.	0-10

Table 9-4: Procedural Defenses and Substantive Defenses under the OSHAct.

PROCEDURAL DEFENSE	SUBSTANTIVE DEFENSE
1. Failure of OSHA to conform to constitutional and statutory requirements in conducting inspections of employer worksites.	1. Standard cited, or the GDC, does not apply to the employer or it is pre-empted by a more specifically applicable standard.
2. Improper service of the citation and/or the employer working conditions are covered by another federal agency, not by OSHA.	2. Impossibility of compliance (technically or economically unfeasible).
3. Failure to issue citation with reasonable promptness once the alleged violation is discovered.	3. Unpreventable employee misconduct.
4. Failure of the citation to state the alleged violation with particularity.	4. Implementation of the cited standard would create a greater hazard than the existing alleged non-compliance condition.
5. Citation was issued more than six months from the occurrence of the alleged violation (expiration of the statute of limitations).	5. Failure of OSHA to prove the employer had knowledge of or could have with the exercise of reasonable diligence had knowledge of, the alleged hazardous condition or the standard that led to that condition.

The percentage reduction for citations with high P/S quotients cannot exceed 20 percent. A P/S of 8, for example, will only be allowed a maximum 20 percent reduction regardless of the results of the Table 9-3 review. A P/S of 9 will yield a maximum 10 percent reduction, and no reduction will be permitted for a P/S of 10.

The penalty calculation factors only apply to proposed penalties, which become final only if the employer fails to file a notice of contest. The OSHRC retains final authority to assess penalties. Depending upon the outcome of the contest proceedings, the final penalties imposed may be drastically higher or lower than those proposed by OSHA.

Chapter 10

CONTESTING ALLEGED VIOLATIONS AND PENALTIES

OVERVIEW

This chapter focuses on the employer's right to dispute (or *contest*) any OSHA citations that allege safety violations as well as any proposed penalties that accompany those citations. Employers can do this through *informal settlement* procedures which occur under the authority of the local Area Director. If a settlement can not be reached, then the employer has the option of pursuing *formal settlement* through a hearing before the OSHRC.

Employers can contest the entire citation itself or only certain aspects of it. For example, if the employer can show the abatement period to be unreasonable or impossible to achieve, then a *petition for the modification of abatement (PMA)* can be filed. Employees or their representatives also have limited contest rights under current law. If they disagree with the abatement period established by OSHA in the citation, they can file a contest stating so and their reasons for making such a claim. Currently, there are no other provisions for employee contests under the OSHAct. However, it should be noted that various versions of proposed "OSHA Reform" legislation will likely greatly enhance employee contest rights, if Congress should ever pass such legislation.

Also provided in this chapter will be details on the role of OSHRC in the contest process. The issue of *burden of proof* with regard to the employer as well as OSHA under various circumstances will be discussed. In summary, OSHA will have the burden of proving the violations occurred in all proceedings initiated as a result of the filing of a notice of contest.

With respect to an *affirmative defense,* employers will ordinarily have the burden of proof. An affirmative defense is one in which the employer does not dispute the cite itself, but offers substantial proof to justify reasons for not complying with the cited standard(s). The most common of these defenses can be divided into two categories, *procedural defenses* and *substantive defenses.* Procedural defenses focus on the validity of OSHA's enforcement procedures and the procedures used by OSHRC in contested cases. Substantive defenses deal with the validity and applicability of a particular standard to the facts of the case, the nature of the employer's conduct, and its effect on the safety and health of employees.

In such instances, a *preponderance of evidence* must exist which will support the position of the defense. A preponderance of evidence is defined as the amount of evidence which is sufficient to convince the judge that the facts asserted by a proponent are more likely to be true than false.

The entire OSHRC contest process and the rules that have been established to govern each step will also be discussed in this chapter. From the filing of a notice of contest by an employer (or employee on abatement period issues) through pretrial discovery, the trial itself, and post-trial activities, employers must be aware of the process and their rights along the way.

INTRODUCTION

The subject of citations and penalties was discussed at length in the previous chapter. The fact that employers have the right to file a contest to any citation was also presented in brief. This chapter will focus specifically on the *contest process,* the decision to contest and the potential advantages and disadvantages of filing contests, informal settlements, and the role of OSHRC in the adjudication of contested citations.

The OSHAct specifically provides the cited employer with a right to dispute any or all of the allegations that have been made and documented in the form of a citation. This right to dispute, referred to as a contest, can be done formally or informally. *Informal settlements* can be arranged directly with the OSHA Area Director who issued the citation. The Area Director is authorized to affect the end result of the citation quite substantially. From no action at all to the modification of a single line item in the citation to its complete dismissal, the Area Director can offer the employer a wide range of potentially satisfying compromises to the original conditions of the citation. Regardless of the outcome, however, the informal settlement process is the employer's first chance to address or answer any allegations directly to the OSHA office that is making the claims. If, however, no informal or satisfactory settlement can be reached with the Area Director and the employer wishes to challenge the citation further, then more formal actions are required that involve the direct participation of OSHRC for resolution. The contest process flow is outlined in Figure 10-1.

THE DECISION TO CONTEST CITATIONS

Employers have fifteen working days from receipt of the citation to contest it. The Area Office establishes the date the citation was received by retaining the postal return receipt signed by the employer or agent. A contest filed more than fifteen days after receipt of the citation will *not* be considered.

Advantages of Contesting

Among the reasons employers contest citations are:

- The possibility of dismissal of the citation;
- Reduction of the degree of violation (e.g., from "serious" to a lesser offense such as "other than serious");
- Reduction or elimination of the proposed fine;
- An extension of the time available to abate the cited conditions (however, the use of the contest for the sole purpose of stalling the abatement period is not appropriate and may result in further actions against the employer, depending upon OSHA's view of the situation).

Other factors affect the employer's decision to contest. For example, issuance of a willful citation where a fatality has occurred could cause or support a criminal charge if not contested (see Chapter 11). Since accepting the citation can result in allegations of repeat or even willful citations at other workplaces controlled by the cited employer, the employer must consider the citation's implications beyond the cited location.

Disadvantages of Contesting

There is also a potential for unfavorable outcomes resulting from contesting an OSHA citation. These should be balanced against the potential benefits discussed above when deciding whether or not to file contest. Potential disadvantages include:

- Costly legal counsel is often required to contest;
- In accident situations involving injuries, contesting could result in other legal liabilities (depending on the circumstances and the state having jurisdiction, the contest could hinder or support tort litigation).

Initiating the Contest

To initiate a contest, the employer must send a letter to the OSHA Area Director who issued the citation within the fifteen day contest period. The letter should:

- Be sent via certified or registered mail for your protection;
- Identify the employer, the citation being contested, and the issues the employer is contesting;
- Be signed by an authorized individual (such as a plant manager or owner).

Consequences of Not Contesting a Citation

If an employer chooses not to file a contest within the fifteen day period, then the citation becomes a *final order* of OSHRC and cannot be appealed. Violative conditions must be abated during the time period established in the citation, and any fine paid. The violations can also result in follow-up inspections to confirm abatement of the violation and may lead to either repeat or additional violations and more fines.

Figure 10-1: The process flow for contesting an OSHA citation.

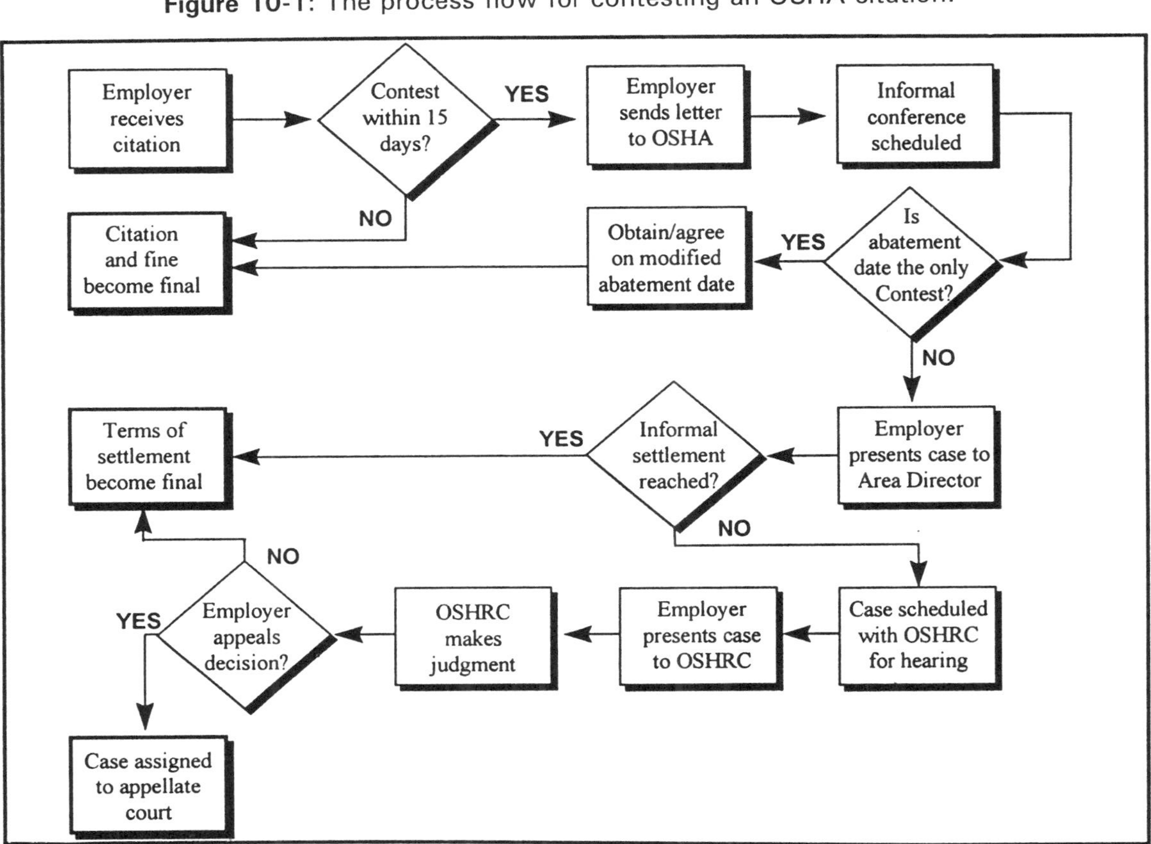

Informal Settlement of Citations

OSHA 29 CFR 1903.19 allows for informal settlement conferences between OSHA, the employer and employee representatives. All parties to the citation are invited to attend the settlement conference. OSHA Area Directors' settlement authority includes the option to reclassify alleged violations, or to change or delete a penalty, citation, or individual items in the citation. Informal conferences are always granted before the expiration of the fifteen day contest period. Settlement discussions can continue with the Area Director after a contest is filed, but filing the contest is necessary to preserve employer rights and to continue discussion once the 15-day period has elapsed.

Informal conferences can result in rapid settlement of a citation, especially when the employer can prove that the citation is clearly erroneous or when the employer is abating the cited hazards, but wants to reduce the penalty and/or violation category.

Extending Abatement Time

In order to increase the amount of time for abatement, employers can contest that provision of the citation or file a Petition for Modification of Abatement (PMA).

Abatement Through Contest

Since the time allocated for abatement is a critical aspect of the citation, the time period to abate is automatically contested when the overall citation is contested. However, the contest notice should specifically refute the abatement period as unreasonable or impossible to achieve regardless of the employer's good faith efforts to comply. And, since a challenge to the abatement period alone permits OSHA to consider such a contest as a Petition for Modification of Abatement period rather than an actual citation contest, the contest should challenge other elements of the citation as well.

The employer should note that in the contest procedure, OSHA must prove that the abatement period is reasonable and the opposite is true under the procedure for a PMA as described below.

Petition for Modification of Abatement (PMA)

The PMA procedure (29 CFR 2200.37) establishes an employer's right to petition for an extension of the time allowed for abatement, but the petition must be filed by the working day following the date set for abatement. In rare instances, it can be filed later but only if accompanied by a statement of exceptional circumstances explaining the reasons for the delay. A copy of the PMA must be posted in the workplace where all affected employees can read it.

The PMA must include at least the following specific information before it will be considered:

- Any abatement measures already taken, including their implementation dates;
- How much additional time will be needed to complete abatement;
- The reasons additional time is needed;
- What measures are being used to protect employees while abatement is undertaken;
- Certification of proper PMA posting.

When completing the PMA, it may be helpful for the employer to note that OSHA will usually allow an extension of the period for abatement for either technical or financial reasons. Technical reasons require evidence that the abatement period is not achievable (because it would take longer than the time allowed or because technology to abate the condition is not available). Since proposed abatement methods must be economically feasible, the time period can be extended on financial grounds when abatement poses a significant financial hardship to the employer.

Dealing with the Occupational Safety and Health Review Commission

Once the employer has filed a notice of contest with the OSHA Area Office, the Area Director has 15 working days to send the notice and all associated documentation to the Executive Secretary of the Occupational Safety and Health Review Commission. Once this is accomplished, jurisdiction over the proceeding will rest solely with the Commission.

As discussed in Chapter 2, OSHRC was created as an independent agency to handle adjudication under the OSHAct. The Commission reviews citations and penalties contested by the cited employer. Employees or their representatives may only contest a citation's proposed abatement dates. Under current provisions, employees do *not* have rights to contest any other element of a citation. Many proposed versions of what has been commonly referred to as "OSHA Reform Legislation" attempt to change this aspect of the contest process and give employees broader contest rights with respect to citations.

The OSHRC has many of the same powers of federal courts, including the power to promulgate rules of practice, issue subpoenas and compel testimony, issue orders based on evidence and legal principles and set civil penalties. The President appoints all three Commissioners with the advice and consent of the Senate, and chooses which member will serve as Chairman. The Commissioner's length of term is six years, except when filling a vacancy (due to death, resignation, or removal from office prior to term expiration), when the replacement Commissioner only serves the remainder of the term.

The OSHRC Chairman appoints administrative law judges (ALJs) and other employees needed for Commission operations. The ALJs conduct all pre-hearing and hearing matters once a case has been sent to the Commission. To qualify as an ALJ, candidates must have been practicing attorneys for a minimum of seven years and must pass the Office of Personnel Management screening process.

The ALJ's are responsible for:

- Conducting fair and impartial hearings;
- Rendering decisions swiftly;
- Issuing subpoenas;
- Ordering depositions;
- Handling pleadings, motions, and other procedural matters; and
- Calling and examining witnesses.

The ALJs usually conduct hearings where the alleged violation occurred. Employers are not expected to travel to Washington, DC or some Regional Office to be heard. ALJs' decisions become OSHRC final orders unless a Commission member orders the decision be reviewed within 30 days of filing the decision.

THE BURDEN OF PROOF ISSUE

In most contest hearings, including all employer contests as well as any employee challenges to the reasonableness of the abatement date, it is OSHA that must prove that a violation has occurred. The exceptions are PMA proceedings and affirmative defenses, where the employer (the "petitioner"), not OSHA, has the burden of proving that the allocated time is inappropriate.

The Affirmative Defense

Employers normally have the burden of proof for an *affirmative defense,* one in which the employer is not specifically arguing the fact that the cited condition(s) existed. The defense is really to the contrary. By not disputing the citation itself, the employer actually agrees with (affirms) the allegation of non-compliance but offers substantial proof to justify reasons for not complying with the cited standards. The contest in such cases can be based upon a number of factual circumstances that made compliance either impossible, impractical, or even more dangerous than the cited non compliant condition. The employer takes up a defense to justify the contest. The most common of these defenses can be divided into two categories, *procedural defense* and *substantive defenses,* as shown in Table 9-4. Procedural defenses focus on the validity of the Secretary's enforcement procedures and the procedures used by the OSHRC in contested cases. Substantive defenses deal with the validity and applicability of a particular standard to the facts of the case, the nature of the employer's conduct, and its effect on the safety and health of employees.

In both cases, the employer must substantiate or "prove" their defense with a preponderance of evidence.

The Preponderance of Evidence Requirement

The *preponderance of evidence* test, well known in civil litigation, applies in OSHRC proceedings as well. According to the Commission, a preponderance of evidence is defined as that quantum of evidence which is sufficient to convince the judge that the facts asserted by a proponent are more probably true than they are false. In most situations, including willful, repeated, serious, other than serious, or failure to abate violations, it is the responsibility of OSHA to provide such evidence when an employer files a contest. Exceptions are in those instances where the employer disputes the abatement period or is planning an affirmative defense. In such cases, it is the employer who must establish a preponderance of evidence to support those positions.

General Proof Requirements

Employer's responsibilities for workplace safety and health derive from the general duty clause, the employer's duty to comply with OSHA standards.

For violations of the general duty clause, OSHA must prove by a preponderance of evidence that:

- The employer did not eliminate the cited hazard at the cited workplace;
- Said hazard was known to the employer or was common knowledge in the employer's industry;
- Said hazard caused or was likely to cause death or serious physical harm; and
- Feasible means were available for the employer to eliminate or control the hazard.

For violations of OSHA standards, the agency must show by a preponderance of evidence that:

- The cited standard is applicable to the employer;
- The employer did not meet standard's requirements;
- Employees were exposed to the hazard covered by the standard; and
- The employer either recognized the hazard or could have known of it through the exercise of reasonable diligence.

Other Proof Requirements

The Commission and the courts have developed other proof requirements for OSHA enforcement cases. In Chapter 1, it was established that OSHA must prove that cited employers are *engaged in interstate commerce.* Since employers will simply admit to it during the proceedings, OSHA rarely needs to supply such proof. But if an employer is not engaged in interstate commerce and makes this part of the challenge, OSHA would have to prove the agency's jurisdiction over the employer.

Technological and Economic Feasibility

In cases regarding violations of standards regulating chemical or physical hazards, OSHA must prove both *technological* and *economic feasibility* of the proposed abatement methods. When the abatement period is stated as "immediate," OSHA must show with a preponderance of evidence that this is "reasonable." Since safety issues can often present an immediate hazard which must be corrected without delay to avoid employee injury, OSHA can therefore justify the immediate abatement requirement quite easily. But many health cases include a challenge of the reasonableness of the abatement date because while immediate onset of illness can occur, it is more probable that employee exposures will not "immediately" result in illness.

Also, reasonable abatement time in such cases must take into consideration the feasibility of the abatement strategy. It is not uncommon to encounter abatement methods to correct health hazards which are both technologically and economically infeasible. OSHA may have difficulty establishing a preponderance of evidence in such cases. These issues must therefore be considered carefully by an employer who is contemplating a contest to an abatement period for health-related citations.

And, in many cases, especially general duty clause cases, OSHA must demonstrate how implementation of their proposed abatement strategy will end or significantly control the allegedly hazardous condition or practice. In other words, it is not enough for OSHA to prove that its proposed abatement methods are economically and technologically feasible. OSHA must also demonstrate a reasonable assurance that there will be an appreciable and corresponding improvement in working conditions.

THE EMPLOYER'S REBUTTAL CASE

Employers should present *all* available defenses when providing rebuttal in answer, to the OSHRC during contest proceedings. This should be done as early as possible or the OSHRC may determine that the employer has waived their rights to such defenses, thereby making them unavailable from that point forward.

As discussed above, the employer can utilize an array of defenses which are either *procedural* or *substantive* (Table 9-4). Procedural defenses rely primarily on the possibility that OSHA or OSHRC made some error in inspection, citation, or during contest proceedings, the nature of which establish a legitimate defense for the employer. There are also numerous substantive defenses available to the employer. Employers considering a challenge should choose the defense most applicable to the citation in question.

Table 9-4 lists the most common substantive defenses encountered. Of these, the following three are perhaps the most recognized and well-developed.

Impossibility of Compliance

This is a defense that is commonly used by employers who cannot directly challenge OSHA's case based upon the established facts. Employers can avoid liability for non-compliance if they can prove compliance would have been impossible because such compliance would prevent the employer from continuing its operations. The employer must prove that compliance was economically infeasible or impossible and that other protection options were either in use or not available. Since difficulty of compliance rather than pure impossibility is usually an issue, OSHRC has generally not ruled in favor of this defense tactic.

Unpreventable Employee Misconduct

As discussed in Chapter 6 (Employer Compliance), this is another commonly encountered substantive defense. With this defense, the employer must show that it has implemented work rules intended to prevent OSHA violations, and that the rules have been adequately communicated to employees. Also, the employer must show that it has used an inspection program or some other means to discover violations and that its safety and health rules have been enforced when violations occur.

The Greater Hazard Defense

OSHRC sometimes lets employers avoid sanctions for violations of otherwise applicable safety regulations if abating the hazard would itself be a worse threat to employees' safety and health. This defense requires employers to prove that the hazards of compliance exceed the hazards of noncompliance, that there are other employee protection options, and that a variance to the cited standard could not be secured.

Any employer contemplating a challenge to an OSHA citation should research the great body of case law on this subject and make use of the best affirmative defenses available under the particular circumstances of their case.

OSHRC Hearing Procedures and Rules

OSHRC's procedural rules (29 CFR 2200.1-211) dictate that as soon as the Commission receives an employer's Notice of Contest and associated documents, the case is assigned a docket number and the employer, the Department of Labor, and any interested third parties (employees, labor unions, etc.), are notified of the case and docket number. OSHRC also requires employee notification regarding the content and furnishes employers with a standard notice form to be delivered to employees along with a return post card to prove that the employees have been notified. This process is explained in a packet containing a copy of 29CFR 2200.1-211 sent to the employer by the Commission. It should be noted here that failure to properly notify employees can result in dismissal of the notice of contest.

The OSHRC procedures for preparation and trial of OSHA cases cover:

- When and how various notices to the affected parties (employers, employees, employee representatives, etc.) will be delivered;
- The use of Federal Rules of Evidence for all hearings;
- Case pleadings;
- The discovery process (depositions, interrogatories);
- Scheduling and conduct of hearings;
- Commission review of ALJ decisions; and
- Settlement orders.

If no OSHRC rule applies in a particular case, proceedings are held in accordance with the Federal Rules of Civil Procedures.

Two other rules issued by the commission establish expedited and simplified proceedings. OSHRC can order an *expedited* proceeding, if requested by any party or by intervening in the case. Employee contests (regarding abatement dates) or petitions for the modification of abatement periods are expedited automatically under these rules. Expedited proceedings move ahead of all other types of cases on the ALJ's and OSHRC's review dockets, and filing deadlines are replaced with an expedited schedule established by the ALJ or Commissioner.

Simplified proceedings only apply to certain types of cases (they may not be used in GDC cases or those alleging violations of health, noise, vibration, or radiation standards, Commission Rule 202). The employer or employee representative can file a request for simplified proceedings within 10 calendar days after the docket number and other documentation has been received from OSHRC. Parties have 15 days after receipt of the request to file an objection, which will cause the request for a simplified proceeding to be denied. If no objections are received, simplified proceedings will be initiated, meaning no pleadings, discovery, interlocutory appeals, or motions that have not been first reviewed by all parties prior to the hearing. In addition, the parties must discuss settlement, clarification of issues, and other such matters before the conference or hearing is convened.

A simplified proceeding *conference*, presided over by an ALJ, is held to try to clarify unresolved issues. A *hearing* then addresses these unresolved issues, after which the ALJ issues a written decision. Hearings follow the Administrative Procedures Act (APA) format and the same review procedures apply, but the Federal Rules of Evidence do not.

Pleadings and Motions

Pleadings are the statements setting forth to the court the claims of OSHA (the plaintiff) and the answers of the employer (the defendant). The OSHRC has very detailed rules on the form of pleadings. For example, rule 30 requires all pleadings and documents (including any exhibits) to be typed, doubled-spaced, on 8-1/2" x 11" opaque paper, with one and one-half inch margins, and fastened in the upper left hand corner. Rule 31 requires the proper use of caption and title of cases and placement of the docket number on the first page of every pleading or document (except exhibits). Other OSHRC rules, such as those on the service of pleadings, deadline calculation, and requests for extensions, should also be consulted prior to filing.

Motions are simply applications to the court for a ruling or order. Motions are governed by rule 40, which establishes that requests for orders shall be made by motion, and not as part of some other pleading such as a complaint or answer. Barring an ALJ's orders to the contrary, all motions are made in writing. In urgent circumstances, however, where the situation calls for immediate action or attention, a motion may be made by telephone if it is filed in writing shortly thereafter. Any party opposing a motion has ten days from service of the motion to file a response. Motions to dismiss, for a more definitive statement, for summary judgment, and motions for consolidation are the most common.

Pretrial Discovery

The OSHRC's rules of discovery preempt the Federal Rules of Civil Procedure, except in the absence of specific OSHRC provisions. OSHRC's primary means of discovery are the following:

Production of Documents. Requests for the production of documents (or other such materials), or permission to enter upon land or other property for inspection and other purposes (Rule 53). These requests can only be made after the filing of the first responsive pleading or motion that delays the filing of an answer (such as a motion to dismiss), so discovery cannot commence until the answer is filed. The request should clearly describe the specific items and/or establish a time for the inspection to take place, and the opposition must respond within thirty days of receipt of said request.

Admissions. Requests for admissions can also only be served after the first responsive pleading or motion that delays the filing of an answer has been filed. An admission is essentially a granting of truth or a conceding of truth from the courts. The maximum number of requested admissions is 25, including subparts, unless the judge grants more. The request for a higher number should be made by a motion indicating why more than 25 requests for admission are required. Within 30 days of receipt, if no objection is received from the opposing side, the requested admission will be admitted.

Interrogatories. As many as 25 interrogatories can be served by each party without permission from the court after the first responsive pleading or motion that delays the filing of an answer is filed. Interrogatories are nothing more than formal sets of questions, usually written, specific to the case, that must be answered by the party served, usually in writing. If more than 25 are required, a party must file a motion detailing whether the complexity of the case or the number of cited items creates a need for a higher number of interrogatories. Interrogatories within interrogatories are counted as separate interrogatories, all of which must be answered within 30 days.

Depositions. In addition, depositions are permitted under limited circumstances. To take depositions, the requesting party must obtain the agreement of all parties involved or seek an OSHRC or ALJ order following filing of a motion stating *good and just* reasons for granting the request. As in the case of the other discovery means discussed above, the motion to take depositions can be filed only after the filing of the first responsive pleading or motion delaying answer. Parties to be deposed must receive ten days written notice, except if the parties have waived this requirement. Any expenses related to or incurred as a result of the deposition must be borne by the party who requested the deposition. Depositions may be used as discovery, to counter the testimony of others, or for other purposes listed in the Federal Rules of Evidence or the Federal Rules of Civil Procedure.

Subpoenas. Litigants in OSHRC proceedings also have the right to use subpoenas to force witnesses to testify and/or produce evidence. Applications for subpoenas must be filed either with the judge after the case has been assigned, or with the Executive Secretary of the Commission before the case has been assigned. Parties served a subpoena have 5 days to oppose it. Opposition can be on the grounds that:

- The evidence that will be produced is irrelevant to the issues being investigated; or
- The subpoena does not describe in sufficient detail the evidence that must be produced; or
- Any reason under the law making the subpoena invalid.

There are also rules which allow sanctions for failure to make discovery. Essentially, these rules expressly permit the judge to enter an order dismissing the action proceeding or any part thereof, or rendering a judgment by default against the disobedient party. In other words, a party can lose its case by default because of a failure to make pretrial discovery.

The Trial

The OSHAct requires that the employer be granted a hearing whenever a citation or a notification of penalty for failure to abate has been challenged. This formal administrative hearing, similar to a civil trial without a jury, is conducted under the Administrative Procedures Act (APA). Witnesses testify under oath and are cross-examined, and documentary evidence is received. As discussed previously, the Commission's rules govern the conduct of its hearings, with the Federal Rules of Civil Procedure in force where OSHRC rules are silent.

Written notice of a hearing date and location is provided to all parties at least thirty days in advance of the hearing, unless the hearing has been previously postponed. In the latter case, notice will be given 10 days in advance of the hearing. Employers are then required to notify all affected employees of the hearing date and location.

Hearings can be postponed upon the motion of an affected party or at the ALJ's discretion. Any motion must detail the other parties' position, the reasons why the stay is requested and the proposed length of stay. The motion must be filed by the seventh day before the hearing (unless good cause can be shown to justify shorter notice). However, it should be noted that *stays* of proceedings (delay or postponement) are only granted occasionally, and are not looked upon favorably by OSHRC.

The Judge's Decision

The presiding judge will generally not issue an immediate decision at the end of the hearing. Normally, the ALJ allows the parties time to mail briefs presenting their positions by a pre-determined date, typically from one to three months after trial. The judge may then take an additional month or more to review these materials before ruling on the case.

The judge must issue a written decision covering findings of fact, conclusions of law, and the reasons or bases for them, on all disputed matters. An *order* upholding, modifying, or vacating each contested citation item and proposed penalty will be included. If a petition for modification of abatement period is involved, the decision will include an order supporting or modifying the original abatement period. Once rendered, the decision, unless changed, will be filed within 20 days. The *docketing date* begins the 30-day period during which the Commission may decide to review the decision.

Commission Review

Although rarely granted, parties adversely affected or aggrieved by an ALJ's decision may file a petition for review by the Commission with either the ALJ during the 20 day period before the decision is sent for docketing, or with the Executive Secretary of the OSHRC within the 20 days after receipt of the decision. An *aggrieved* party must file for review in order to preserve its right to appellate court.

On rare occasions, an OSHRC Commissioner chooses to review a decision after it is docketed. If the case includes unique matters of law or policy, the Commission may decide to review the decision.

If no Commission member directs review of the ALJ's decision during the 30 days following docketry, then that decision becomes OSHRC's final order. If it is to be reviewed, the Commission will request briefs from parties before deliberations and may grant motions for oral arguments if they would facilitate its decision-making process. Oral arguments are presented at the Commission's Washington, D.C. office.

While the Commission normally supports the ALJ's decisions, it can overturn them, establishing its own findings of fact as well as reasons for the reversal.

Appealing to the Federal Courts

Any party who has lost its case before the OSHRC can appeal the decision to the United States Court of Appeals for the circuit in which the alleged violation occurred, where the employer's main office is located, or with the Court of Appeals for the District of Columbia. OSHA appeals must be filed in either of the first two locations available to the employer.

A three judge panel will base its decision on the written transcript, briefs filed by the parties, and oral arguments of under 15 minutes per side. OSHRC's findings are rarely overturned by the appeals courts.

The issues considered by the court are relatively narrow as compared to those of a lesser court. Only arguments made while the case was before the Commission will be considered, and the case will only be reviewed if OSHRC denied a petition review of the decision. Another important aspect of the federal appeals process is the view the courts take on evidence in the case. The OSHAct holds that OSHRC's findings regarding questions of fact supported by *substantial evidence* on the record are considered conclusive. "Substantial evidence" is defined as the relevant evidence a reasonable mind might need to support a conclusion.

OSHRC penalty assessments are reviewed under the *abuse of discretion* standard, which is even more deferential than the substantial evidence standard. Appeals courts are unlikely to change a penalty imposed by the Commission if OSHRC is within its authority and has based its decision on clearly enumerated findings.

Disagreements between OSHRC and OSHA regarding interpretation of an OSHA standard may reach the federal courts. Until 1991, the federal courts did not favor OSHA or OSHRC in these disputes, but the Supreme Court then ruled that OSHA's interpretation was primary as OSHA interpretive authority over standards.

It should also be noted that when an employer appeals to federal courts, a separate request for a stay of the obligations imposed by the Commission's decision must be filed with OSHRC or the federal court or the employer must comply with the order in the interim.

Equal Access to Justice Award (EAJA)

When a private party has prevailed against OSHA in actions before the OSHRC, they may be entitled to compensation for attorney's fees and related legal expenses under the EAJA, unless OSHA's position is considered "substantially justified" or other circumstances exist. To receive an EAJA, the party to an OSHRC proceeding must be one of the following types of business or government entities:

- A sole proprietorship, partnership, corporation, association, government entity, or public or private organization with a net work under $7 million and employing fewer than 500 employees; or

- A cooperative association or a charitable organization with fewer than 500 employees; or
- An individual whose net worth does not exceed $2 million.

As noted above, an EAJA may not be possible if facts exist to substantially justify OSHA's position. Substantial justification must be based on a the merits of the case, the governmental action (or lack thereof) that the case is based on, and the government's position.

To obtain the award, the party must file an application within 30 days following OSHRC's final action. The amount of the EAJA is based on attorneys', agents', and expert witnesses' normal fees, even if the services were pro bono or at a reduced rate. However, the maximum amount payable for attorney's fees is $75.00 per hour.

Chapter 11

CRIMINAL PROSECUTIONS

OVERVIEW

There are four clear provisions in the OSHAct stipulating the circumstances under which employers can be found in criminal violation. The first two are quite obvious. Providing *advance notice* of an inspection and *murdering* a Department of Labor representative are clearly criminal offenses punishable by a number of years in prison and/or monetary fines. The third criminal offense discussed in the OSHAct pertains to the giving of false statements or *false reporting* to OSHA. The fourth criminal offense if *willfully violating* a standard that led to the *fatal injury* of at least one employee.

Through the past twenty-five years, OSHA's use of the criminal violation provisions of the OSHAct has been very infrequent. However, employers should be aware that such provisions do exist and can be used to impose prison time and/or monetary fines against corporations and individuals if the circumstances permit. Also, while individuals have a right to escape self-incrimination under the Fifth Amendment of the Constitution, corporations have no similar rights. This means that company officers and officials can be compelled to testify against their own corporation as well as fellow managers during trial.

This chapter will provide a brief discussion on the subject of criminal prosecutions under the OSHAct as well as the rights of individuals and companies when faced with such actions.

INTRODUCTION

Several chapters throughout this text have contained reference to the fact that criminal prosecutions are possible under the OSHAct. While it was never the primary intent or purpose of Congress to seek such actions against employers, it was understood from the beginning that willful, callous or otherwise purposeful acts of disregard for employee safety and/or health qualified as criminal acts and should be prosecuted as such. The OSHAct, therefore, contains four distinct provisions which define crimes. There are also levels of crime, such as misdemeanors and felonies, which can be prosecuted under the OSHAct.

FEDERAL CRIMINAL PROSECUTION

Section 17 of the OSHAct contains the four provisions which define crimes. The first two are self-explanatory and will not be discussed at great length in this chapter (see Figure 11-1). They are as follows:

- It is a misdemeanor, punishable by imprisonment of up to six months, a fine of $1,000, or both, to provide advance notice of an OSHA inspection;
- The murder or attempted murder of any Labor Department law enforcement investigator is punishable by imprisonment up to and including a life sentence.

The other two crimes under the OSHAct involve false reporting of information to OSHA and willful violations of standards and regulations resulting in one or more fatalities. But these prosecutions are rare. In fact, of the approximately 87 criminal cases that OSHA has referred to the Department of Justice (DOJ) from 1971 to 1993, only 26 have been prosecuted. This has been due in large part to DOJ's reluctance to try misdemeanor criminal safety and health cases. Attorneys must spend nearly the same level of time and resources prosecuting these cases as they do for other criminal cases where the potential results are much more meaningful in terms of the effect on the guilty party (i.e. felonies).

But OSHA has not explained why it has referred so few cases to the DOJ for prosecution. The statistics concerning workplace and job related fatalities bare sharp contrast to this low number of criminal prosecutions. NIOSH has reported that the number of workplace fatalities average 7,000 per year.

Figure 11-1: The four types of criminal activities, as defined in Section 17 of the OSHAct.

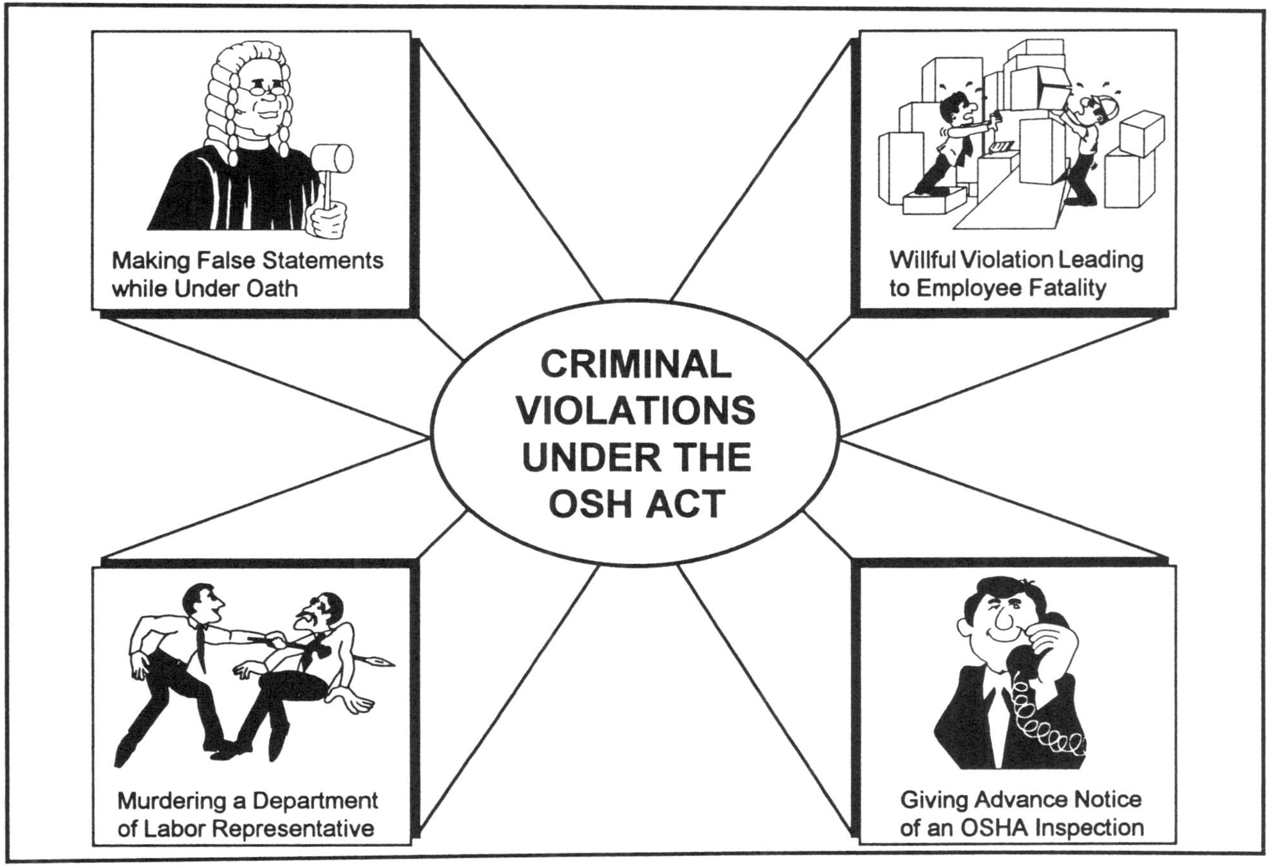

CRIMINAL VIOLATIONS UNDER THE OSH ACT

Making False Statements while Under Oath

Willful Violation Leading to Employee Fatality

Murdering a Department of Labor Representative

Giving Advance Notice of an OSHA Inspection

One reason for the low number of criminal case referrals may be that the OSHAct is a civil statute under which employers are expected to voluntarily eliminate hazards instead of contesting citations. Therefore, OSHA was not expected to utilize DOJ's resources with regularity. Since criminal cases are almost always contested, these facts might imply that OSHA purposely does not refer criminal cases so as to avoid an expenditure of resources. However, it is not the purpose or intent of this text to imply or suggest anything. This is merely a presentation of the known facts concerning criminal referrals and readers are left to form their own conclusions.

A Willful Criminal Violation

The original language of the OSHAct stipulated quite clearly that any employer who willfully violates any standard or regulation pursuant to the Act that causes the death of an employee shall, upon conviction, be punished by a fine not to exceed $10,000, imprisonment for more than six months, or both. If the conviction is for a violation committed after a prior conviction, then punishment escalates to $20,000 and/or imprisonment for not more than one year. Amendments to Title 18 of the U.S. Code have increased the fines to $250,000 for individuals and $500,000 for corporations.

The language of this section of the OSHAct can be broken down to the following three essential elements of a criminal violation that highlight the major difference between civil and criminal prosecutions under the Act:

- An OSHA standard must have been violated;
- The violation had to be committed willfully;
- One or more employees died because of the violation.

Another important difference compared to civil cases is that in criminal cases, prosecutors must prove their cases *beyond a reasonable doubt.* Criminal cases are tried by juries with a much higher *burden of proof* than in civil cases. Construction safety standards, and in particular trenching operation rules, have resulted in the most successful criminal actions. This is most probably because the issue of whether the violation caused the fatality is rather clear in these cases, leaving only the question of whether the violation was willful.

Some courts have determined that a violation is criminally willful if it occurred through the employer's *indifference* to its compliance obligations, while others have not held that indifference supports a criminal finding. Normally, it would be beneficial to examine the "typical case" to help understand and resolve this issue. However, as mentioned previously, criminal referrals of OSHA cases are few and far between and there has simply not been enough case law to establish what a typical case might be. In some instances, employers have received previous civil citations, allowing the prosecutors in the criminal case to prove employer knowledge of its compliance obligations, making subsequent violations willful.

In still other cases, the courts have looked to individuals within corporations liable for the crime and have sought prosecution against them. This is in contrast to the wording of the OSHAct itself, which states that the employer is the party responsible for compliance. Employers are traditionally held responsible for committing violations, not their agents or managers. Appellate courts have decided that it was not Congress' intent that employees be punished under the OSHAct, and that it would be "double counting" to hold employers and employees alike liable for the same violations. But a word of caution is warranted here. The courts have also concluded that *senior corporate officers* and *directors* can be held liable as employers, since they are responsible for charting the business course of their organization and making the decisions necessary to run the enterprise on a day-to-day basis.

Making False Statements

It is a crime under the OSHAct to make false statements to the Department of Labor. Anyone knowingly making a false statement will, upon conviction, be punished by a fine of up to $10,000 and/or imprisonment for up to six months. Therefore, false statements of any kind, including those made in representation, certification in application, record, report, plan, or other document filed or required to be maintained under the OSHAct are a criminal violation of the Act.

This provision also has not been used frequently. The false statement provision does not require a workplace fatality as a condition of conviction. For instance, OSHA standards often include requirements for employers to generate and maintain records (Hazard Communication and Bloodborne Pathogens, for example) or to make written certifications as to the status of their compliance program (Lockout/Tagout). Any knowing misrepresentations or misconduct in recordkeeping is enough for a criminal conviction under this provision of the OSHAct.

THE RIGHTS OF COMPANIES AND INDIVIDUALS

Although corporations have a Fourth Amendment right of privacy, they have no Fifth Amendment right against self-incrimination. So any statements made by corporate managers or other officers can be used to convict their employer companies of OSHAct criminal violations. In fact, under OSHA's subpoena authority, corporate managers and supervisors can be compelled to testify and surrender company documents.

However, documents prepared by counsel under the attorney/client privilege or the work product privilege are exempt from discovery. The attorney/client privilege protects communication between those parties, while the work product privilege applies to documents and other work performed by the attorney or under the attorney's direction or control. It should be understood that use of either privilege does not always mean that the materials and documents cannot be obtained. The rules and the case law do permit access to these materials under certain exceptional circumstances. However, in the typical case, it is not the application of the rules which cause privileged materials to be discovered. It is more common to find that employers will inadvertently or through mistake or ignorance create a waiver to the privilege. For example, a mistaken publication of a privileged document can waive the privilege and open it to discovery. Employers should therefore be well versed in the requirements of attorney/client privilege and work product privilege before careless mistakes create embarrassing or incriminating legal problems during litigation.

Although a corporate manager can be compelled to testify by subpoena, he or she has the right to refuse to testify to avoid self-incrimination. But the Fifth Amendment cannot be used to protect the employer, so managers can be compelled to testify against their colleagues. Some prosecutors will even grant immunity to some defendants to insure their cooperation in convicting defendants.

Federal Preemption of State Prosecutions

While both federal OSHA and local enforcement agencies have jurisdiction to prosecute for violations under their respective statutory authorities, there has been some question whether prosecution for federal civil and criminal violations, as well as prosecution by the state and local authorities, would amount to double jeopardy under the Constitution. The Supreme Court has determined that the states can prosecute violators of their criminal laws regardless of federal regulations.

Employers, like individual citizens, have an obligation to comply with both federal and local statutes which are applicable to their activities. A major difference between civil and criminal prosecutions is that a criminal prosecution does not result in an order requiring the employer to abate the alleged hazard, whereas, a civil prosecution does result in an abatement order. This may be enough of a distinction for the courts to allow prosecution of the civil case because it seems reasonable to assume that an employer faced with a criminal prosecution will attempt to have the civil case stayed while the criminal charges are pending. The result of such strategy will theoretically delay the abatement of the alleged hazard and defeat the intent of the OSHAct in the process. Clearly, the courts would most likely utilize their authorities to prevent such an outcome.

State and Local Prosecutions

It is a fact that some state and local authorities have prosecuted employers under state criminal statutes, primarily when workplace fatalities have occurred. The statutes used to prosecute include criminal assault, criminal negligence, manslaughter, and murder. State and local prosecutions also vary widely and are largely dependent on the desire of the local prosecutors to press for criminal charges when employees are injured or killed on the job.

Chapter 12

OSHA AND CIVIL SUITS

OVERVIEW

While it is a fact that the OSHAct was never intended as a replacement for the various state worker's compensation laws, injured workers have successfully used citations for OSHA violations as proof of negligence in civil suits. In fact, OSHA violations have been used in private litigation to establish the negligence of the defendant. In general, a plaintiff suing under negligence theories must show that the defendant owed a *duty of care* to the plaintiff, that the duty was somehow unfulfilled ("breached") by the negligent action or inaction of the defendant, and that such action or inaction was the proximate cause of the plaintiff's injury.

The plaintiff must also substantiate that damages claimed are in some fashion due to the defendant's negligence. Hence, proof of the violation is used to eliminate the burden of establishing negligence, leaving the plaintiff (the employee) with the burden of establishing the duty of care, proximate cause, and resulting damages. In other words, proof of an OSHA violation can be used to help the employee over the first crucial hurdle of the burden of proof requirement.

Employees have also been able to circumvent the restrictions imposed by some worker's compensation laws in states which recognize a *dual capacity doctrine*. In essence, if an employer acts in a capacity other than that of employer in a dual capacity state and an employee is injured, then the employer may be sued for negligence arising out of its dual capacity role.

INTRODUCTION

In Chapter 2, the purpose and intent of the OSHAct was introduced and discussed. Throughout this text, the OSHAct has been explained and its provisions have been detailed. Here, in the final chapter, the reader is reminded that the OSHAct was not intended as a means of aiding or assisting private litigants in their pursuit of damages for occupational injuries or illnesses. Nor was it intended as a replacement for worker's compensation, which has been the traditional means (along with the tort law system) of obtaining restitution for such injuries and illnesses. It is true that when the Act was passed, Congress was well aware that the existing worker's compensation system had not produced the desired reduction in workplace injuries and illnesses. But Congress did not at that time intend to displace or replace that system. OSHA was created to coexist with worker's compensation.

But the OSHAct has had an effect on civil litigation and worker's compensation regardless of its intent to do otherwise. This chapter will provide some discussion of the Act's direct and indirect effects on civil litigation.

CIVIL LITIGATION AND THE OSHACT

Injured workers and their surviving family members have successfully sued to recover damages for occupational injuries and illnesses. While the OSHAct does not permit direct suits against employers, plaintiffs (or more specifically, their attorneys) have used third parties in order to achieve their objectives. For instance, employers can become liable for illnesses and injuries if they accept responsibility for the safety and health of other employers' workers, or under indemnification clauses of work contracts as the contracting agency. Several states' laws cause employers to become liable to their own employees, existing worker's compensation laws notwithstanding.

Employers are generally liable to their employees for occupational injuries and illnesses, regardless of negligence, under a state's worker's compensation program. The word *negligence* is used because employers are not exempt under worker's compensation for intentional torts. If an employer, with premeditation or afterthought, deliberately exposes its employees to a known toxic substance at levels known to be harmful to workers, an intentional tort and a crime equivalent to assault with a deadly weapon has likely been committed. These employers have been held accountable for their actions in private litigation. In fact, some states, such as Florida and West Virginia, have, through legislative action or judicial decisions, modified their worker's compensation laws, to make willful OSHA violations subject to private litigation.

The Dual Capacity Doctrine

Some states, including California, Illinois, and Michigan, recognize a *dual capacity doctrine*. This doctrine states that an employer can have a dual role (employer and some other entity) where employees are concerned. As an example, assume an employer has employees who are required to work with hazardous materials (not uncommon in today's workplace). While working with these materials, an employee is injured and an in-plant physician provides first aid or other more advanced medical care to the employee (see Figure 12-1). In states with no dual capacity doctrine, the employer would be protected from suit under the state's worker's compensation laws. However, in a dual capacity state, the same employer may not be totally protected under the worker's compensation umbrella. At the moment the employer renders medical services, it may no longer be viewed as the employer exclusively. Rather, the state might consider the employer a *health care provider* instead. If the health care was faulty or otherwise negligent or even suspected of being so, the employee may be able to seek recovery separate from that which is available under the state's worker's compensation remedies.

Negligence and the OSHAct

OSHA violations can be used to establish employer negligence in private litigation. Employees suing under negligence theories must generally prove:

- That the employer owed the employee a duty of care;
- That this duty was somehow unfulfilled ("breached") by the employer's negligent action or inaction;
- That said action or inaction was the cause of the employee's injury.

Figure 12-1: In states with a "dual capacity doctrine," employers can be viewed as serving more than one role and, therefore, be liable for their actions in more than one capacity.

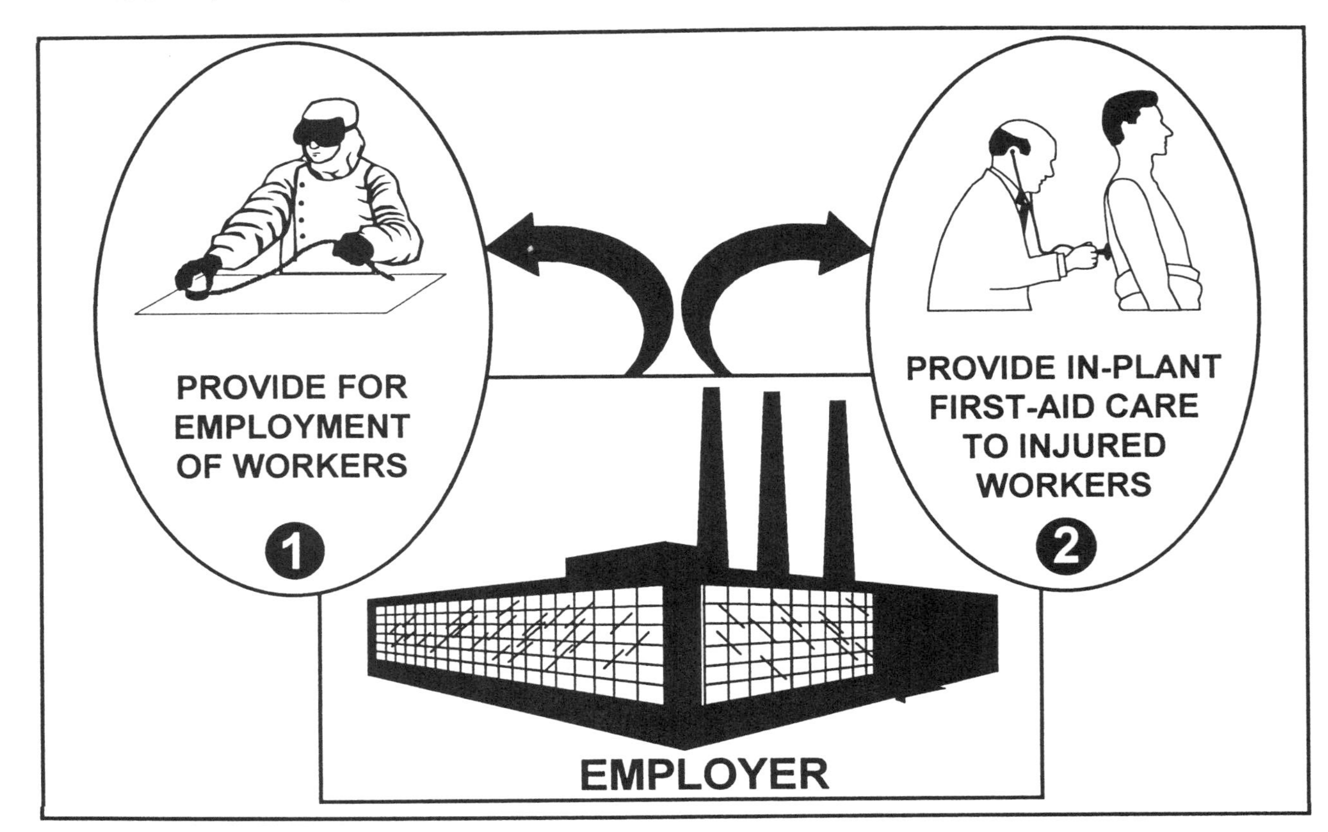

A major element of this equation is that the employee must have been harmed in some way by the employer's negligence. Plaintiffs have at times been successful in using OSHA violations as proof that the employer's actions or lack thereof constituted negligent acts, with proof of the violation eliminating the burden of establishing negligence. The employee then has only to establish the duty of care, proximate cause, and resulting damages, with proof of an OSHA violation establishing the first requirement of proof in a civil case.

OSHA'S EFFECT ON CIVIL PROCEEDINGS

Section 4(b)(4) of the OSHAct stipulates that no provisions of the Act provide injured or ill employees with a private right of action. The courts have held that such a right would allow injured employees more compensation than is allowed under a state's worker's compensation laws. State law generally decides who can be sued for what and under what circumstances and what effect OSHA citations and violations have on civil litigation. Because state laws differ widely, it is not within the context of this chapter to attempt explanation or discussion of state law peculiarities. Refer to individual state laws for clarification.

Injured employees and, more accurately, their attorneys, devise strategies to get around limitations of applicable worker's compensation recoveries. One common way employees overcome such limitations arises in multi-employer worksites, such as construction sites and factories using contract labor. Construction general contractors commonly include indemnity clauses in their contracts with subcontractors. In general, indemnification provisions can effectively release the general contractor from liabilities that may arise as a result of their subcontractor's negligence. In states where the subcontractor's employees can sue the general contractor, the indemnity clause allows them to get to their employer through a legal "back door." The reverse applies if the general contractor provides indemnification to the subcontractor.

In short, employees can use parties other than their own employers who may have owed them a duty of care which may have been breached by an OSHA standard violation, resulting in an injury. If an employee can successfully argue each of these aspects and show that they were damaged (physically, emotionally, financially), then a jury may rule in their favor. In many cases, this is not as difficult a task as it might appear. Insurance companies commonly seek quick settlements. This alone may be enough to prove the first argument that a duty of care was owed as a result of some kind of injury (otherwise, the question may be asked as to why the employer's insurance company was willing to pay if there was no duty owed).

Noncompliance with OSHA as Proof of Negligence

The use of OSHA citations and violations as proof of negligence also varies with state law. In cases where a standard has been violated, it must be determined whether the plaintiff seeking to benefit as a result of the violation was within the classification of persons protected by the standard. For example, a person injured while touring a business establishment is not within the protected class of an OSHA machine guarding standard unless that person is an employee of the establishment.

If the plaintiff is an "employee" who is protected or within the protected class, to what extent the violation of the OSHA standard constitutes negligence must be proven. Under tort law, the employee or plaintiff must prove that the employer failed to exercise the care any reasonable person would have used under the same circumstances. An OSHA standard is seen as a measure of duty owed. Violating a standard shows that the employer did not exercise even the minimum degree of care. Most states agree that proof of an OSHA violation, such as a citation, can constitute evidence of some level of negligence. Violations are therefore admissible as evidence in negligence or tort law cases. In fact, some courts hold that OSHA standard violations are *per se* (by or in themselves) proof of negligence.

When a workplace injury or illness has occurred and a citation is issued, the fact that evidence of a violation could be used as proof of negligence should be considered when deciding to contest. Obviously, such decisions are not easy. But the employer should understand that if it is not contested, the citation becomes final, allowing the citation to stand as evidence of a violation. At the very least, the citation can be used to show proof that the employer somehow neglected its duty to care for its employees, which may then be construed as a probable cause of the injury. The employer's alternative, of course, is to contest the citation, hoping that the OSHRC will vacate it.

Evidence of the circumstances surrounding the injury presented during the contest can be used subsequently by the plaintiff's attorney in a negligence suit. In a settlement of an OSHA challenge, the employer should request that the settlement be worded so that any impact on a negligence case is reduced. And if OSHA alleged a willful violation, the employer should try to have that characterization removed from the settlement agreement.

GOVERNMENT LIABILITY FOR NEGLIGENCE

In some cases, injured employees have sued the federal and state governments, claiming that OSHA inspectors were negligent in enforcing the OSHAct. But suing the government for negligence is not a simple matter. This is because the right to sue the government for negligence is not an absolute right. In the past, government immunity from lawsuits was based on the principle that "the sovereign can do no wrong" (i.e., the government was not liable for its own negligent acts). The Federal Tort Claims Act now requires parties suing the federal government for negligence to state a cause of action under said act. Some states, such as Mississippi, have enacted similar legislation that will permit suits against state governments for negligence.

With respect to OSHA, such suits are usually based on failure to inspect or failure during an inspection. Some suits have been filed based on OSHA's failure to inspect a workplace. However, there is typically little substantiation in such cases to support a suit. Since discretionary acts of government officials are generally immune from suit, OSHA officials' decisions not to inspect a workplace are protected.

Other suits have been based on inspectors' failure to cite a particular hazard during an inspection. In these cases, there may be enough cause to support a suit. It is true that courts have held that inspectors have discretion in deciding whether violative conditions are present. An inspector's decision that no violation exists may be protected as a discretionary act. But if the inspector judged a violation was present but did not issue a citation, a case could be pursued under the Tort Claims Act.

Although subtle, there are important distinctions here. OSHA personnel sometimes disagree about hazards, especially performance standards. For this and other reasons, OSHA often issues compliance directives to assist its staff in consistent administration of agency policies. But field personnel are not always guided by these directives, even thought many of the instructions contained in them may be mandatory, including the issuing of citations under certain violative circumstances. Hence, if a compliance officer fails to cite when citation is mandatory, there may be substantial negligence to support a suit.

Appendix 1

OCCUPATIONAL SAFETY AND HEALTH ACT OF 1970 (PLUS AMENDMENTS)

Public Law 91 - 596, 91st Congress, S. 2193, December 29, 1970
As amended by Public Law 101-552, Section 3101, November 5, 1990

An Act
To assure safe and healthful working conditions for working men and women; by authorizing enforcement of the standards developed under the Act; by assisting and encouraging the States in their efforts to assure safe and healthful working conditions; by providing for research, information, education, and training in the field of occupational safety and health; and for other purposes.

Be it enacted by the Senate and House of Representatives of the United States of America in Congress assembled, That this Act may be cited as the "Occupational Safety and Health Act of 1970".

SECTION 2—CONGRESSIONAL FINDINGS AND PURPOSE

(a) The Congress finds that personal injuries and illnesses arising out of work situations impose a substantial burden upon, and are a hindrance to, interstate commerce in terms of lost production, wage loss, medical expenses, and disability compensation payments.
(b) The Congress declares it to be its purpose and policy, through the exercise of its powers to regulate commerce among the several States and with foreign nations and to provide for the general welfare, to assure so far as possible every working man and woman in the Nation safe and healthful working conditions and to preserve our human resources:

(1) By encouraging employers and employees in their efforts to reduce the number of occupational safety and health hazards at their places of employment, and to stimulate employers and employees to institute new and to perfect existing programs for providing safe and healthful working conditions;

(2) By providing that employers and employees have separate but dependent responsibilities and rights with respect to achieving safe and healthful working conditions;

(3) By authorizing the Secretary of Labor to set mandatory occupational safety and health standards applicable to businesses affecting interstate commerce, and by creating an Occupational Safety and Health Review Commission for carrying out adjudicatory functions under the Act;

(4) By building upon advances already made through employer and employee initiative for providing safe and healthful working conditions;

(5) By providing for research in the field of occupational safety and health, including the psychological factors involved, and by developing innovative methods, techniques, and approaches for dealing with occupational safety and health problems;

(6) By exploring ways to discover latent diseases, establishing causal connections between diseases and work in environmental conditions, and conducting other research relating to health problems, in recognition of the fact that occupational health standards present problems often different from those involved in occupational safety;

(7) by providing medical criteria which will assure insofar as practicable that no employee will suffer diminished health, functional capacity, or life expectancy as a result of his work experience;

(8) By providing for training programs to increase the number and competence of personnel engaged in the field of occupational safety and health;

(9) By providing for the development and promulgation of occupational safety and health standards;

(10) By providing an effective enforcement program which shall include a prohibition against giving advance notice of any inspection and sanctions for any individual violating this prohibition;

(11) By encouraging the States to assume the fullest responsibility for the administration and enforcement of their occupational safety and health laws by providing grants to the States to assist in identifying their needs and responsibilities in the area of occupational safety and health, to develop plans in accordance with the provisions of this Act, to improve the administration and enforcement of State occupational safety and health laws, and to conduct experimental and demonstration projects in connection therewith;

(12) By providing for appropriate reporting procedures with respect to occupational safety and health which procedures will help achieve the objectives of this Act and accurately describe the nature of the occupational safety and health problem;

(13) By encouraging joint labor management efforts to reduce injuries and disease arising out of employment.

SECTION 3—DEFINITIONS

For the purposes of this Act:

(1) The term "Secretary" means the Secretary of Labor.

(2) The term "Commission" means the Occupational Safety and Health Review Commission established under this Act.

(3) The term "commerce" means trade, traffic, commerce, transportation, or communication among the several States, or between a State and any place outside thereof, or within the District of Columbia, or a possession of the United States (other than the Trust Territory of the Pacific Islands), or between points in the same State but through a point outside thereof.

(4) The term "person" means one or more individuals, partnerships, associations, corporations, business trusts, legal representatives, or any organized group of persons.

(5) The term "employer" means a person engaged in a business affecting commerce who has employees, but does not include the United States or any State or political subdivision of a State.

(6) The term "employee" means an employee of an employer who is employed in a business of his employer which affects commerce.

(7) The term "State" includes a State of the United States, the District of Columbia, Puerto Rico, the Virgin Islands, American Samoa, Guam, and the Trust Territory of the Pacific Islands.

(8) The term "occupational safety and health standard" means a standard which requires conditions, or the adoption or use of one or more practices, means, methods, operations, or processes, reasonably necessary or appropriate to provide safe or healthful employment and places of employment.

(9) The term "national consensus standard" means any occupational safety and health standard or modification thereof which (1) has been adopted and promulgated by a nationally recognized standards-producing organization under procedures whereby it can be determined by the Secretary that persons interested and affected by the scope or provisions of the standard have reached substantial agreement on its adoption, (2) was formulated in a manner which afforded an opportunity for diverse views to be considered and (3) has been designated as such a standard by the Secretary, after consultation with other appropriate Federal agencies.

(10) The term "established Federal standard" means any operative occupational safety and health standard established by any agency of the United States and presently in effect, or contained in any Act of Congress in force on the date of enactment of this Act.

(11) The term "Committee" means the National Advisory Committee on Occupational Safety and Health established under this Act.

(12) The term "Director" means the Director of the National Institute for Occupational Safety and Health.

(13) The term "Institute" means the National Institute for Occupational Safety and Health established under this Act.

(14) The term "Workmen's Compensation Commission" means the National Commission on State Workmen's Compensation Laws established under this Act.

SECTION 4—APPLICABILITY OF THIS ACT

(a) This Act shall apply with respect to employment performed in a workplace in a State, the District of Columbia, the Commonwealth of Puerto Rico, the Virgin Islands, American Samoa, Guam, the Trust Territory of the Pacific Islands, Wake Island, Outer Continental Shelf Lands defined in the Outer Continental Shelf Lands Act, Johnston Island, and the Canal Zone. The Secretary of the Interior shall, by regulation, provide for judicial enforcement of this Act by the courts established for areas in which there are no United States district courts having jurisdiction.
(b)(1) Nothing in this Act shall apply to working conditions of employees with respect to which other Federal agencies, and State agencies acting under section 274 of the Atomic Energy Act of 1954, as amended (42 U.S.C. 2021), exercise statutory authority to prescribe or enforce standards or regulations affecting occupational safety or health.

(2) The safety and health standards promulgated under the Act of June 30, 1936, commonly known as the Walsh-Healey Act (41 U.S.C. 35 et seq.), the Service Contract Act of 1965 (41 U.S.C. 351 et seq.), Public Law 91-54, Act of August 9, 1969 (40 U.S.C. 333), Public Law 85-742, Act of August 23, 1958 (33 U.S.C. 941), and the National Foundation on Arts and Humanities Act (20 U.S.C. 951 et seq.) are superseded on the effective date of corresponding standards, promulgated under this Act, which are determined by the Secretary to be more effective. Standards issued under the laws listed in this paragraph and in effect on or after the effective date of this Act shall be deemed to be occupational safety and health standards issued under this Act, as well as under such other Acts.

(3) The Secretary shall, within three years after the effective date of this Act, report to the Congress his recommendations for legislation to avoid unnecessary duplication and to achieve coordination between this Act and other Federal laws.

(4) Nothing in this Act shall be construed to supersede or in any manner affect any workmen's compensation law or to enlarge or diminish or affect in any other manner the common law or statutory rights, duties, or liabilities of employers and employees under any law with respect to injuries, diseases, or death of employees arising out of, or in the course of, employment.

SECTION 5—DUTIES

(a) Each employer:

(1) Shall furnish to each of his employees employment and a place of employment which are free from recognized hazards that are causing or are likely to cause death or serious physical harm to his employees;

(2) Shall comply with occupational safety and health standards promulgated under this Act.

(b) Each employee shall comply with occupational safety and health standards and all rules, regulations, and orders issued pursuant to this Act which are applicable to his own actions and conduct.

SECTION 6—OCCUPATIONAL SAFETY AND HEALTH STANDARDS

(a) Without regard to chapter 5 of title 5, United States Code, or to the other subsections of this section, the Secretary shall, as soon as practicable during the period beginning with the effective date of this Act and ending two years after such date, by rule promulgate as an occupational safety or health standard any national consensus standard, and any established Federal standard, unless he determines that the promulgation of such a standard would not result in improved safety or health for specifically designated employees. In the event of conflict among any such standards, the Secretary shall promulgate the standard which assures the greatest protection of the safety or health of the affected employees.
(b) The Secretary may by rule promulgate, modify, or revoke any occupational safety or health standard in the following manner:

(1) Whenever the Secretary, upon the basis of information submitted to him in writing by an interested person, a representative of any organization of employers or employees, a nationally recognized standards-producing organization, the Secretary of Health, Education, and Welfare, the National Institute for Occupational Safety and Health, or a State or political subdivision, or on the basis of information developed by the Secretary or otherwise available to him, determines that a rule should be promulgated in order to serve the objectives of this Act, the Secretary may request the recommendations of an advisory committee appointed under section 7 of this Act. The Secretary shall provide such an advisory committee with any proposals of his own or of the Secretary of Health,

Education, and Welfare, together with all pertinent factual information developed by the Secretary or the Secretary of Health, Education, and Welfare, or otherwise available, including the results of research, demonstrations, and experiments. An advisory committee shall submit to the Secretary its recommendations regarding the rule to be promulgated within ninety days from the date of its appointment or within such longer or shorter period as may be prescribed by the Secretary, but in no event for a period which is longer than two hundred and seventy days.

(2) The Secretary shall publish a proposed rule promulgating, modifying, or revoking an occupational safety or health standard in the Federal Register and shall afford interested persons a period of thirty days after publication to submit written data or comments. Where an advisory committee is appointed and the Secretary determines that a rule should be issued, he shall publish the proposed rule within sixty days after the submission of the advisory committee's recommendations or the expiration of the period prescribed by the Secretary for such submission.

(3) On or before the last day of the period provided for the submission of written data or comments under paragraph (2), any interested person may file with the Secretary written objections to the proposed rule, stating the grounds therefor and requesting a public hearing on such objections. Within thirty days after the last day for filing such objections, the Secretary shall publish in the Federal Register a notice specifying the occupational safety or health standard to which objections have been filed and a hearing requested, and specifying a time and place for such hearing.

(4) Within sixty days after the expiration of the period provided for the submission of written data or comments under paragraph (2), or within sixty days after the completion of any hearing held under paragraph (3), the Secretary shall issue a rule promulgating, modifying, or revoking an occupational safety or health standard or make a determination that a rule should not be issued. Such a rule may contain a provision delaying its effective date for such period (not in excess of ninety days) as the Secretary determines may be necessary to insure that affected employers and employees will be informed of the existence of the standard and of its terms and that employers affected are given an opportunity to familiarize themselves and their employees with the existence of the requirements of the standard.

(5) The Secretary, in promulgating standards dealing with toxic materials or harmful physical agents under this subsection, shall set the standard which most adequately assures, to the extent feasible, on the basis of the best available evidence, that no employee will suffer material impairment of health or functional capacity even if such employee has regular exposure to the hazard dealt with by such standard for the period of his working life. Development of standards under this subsection shall be based upon research, demonstrations, experiments, and such other information as may be appropriate. In addition to the attainment of the highest degree of health and safety protection for the employee, other considerations shall be the latest available scientific data in the field, the feasibility of the standards, and experience gained under this and other health and safety laws. Whenever practicable, the standard promulgated shall be expressed in terms of objective criteria and of the performance desired.

(6)(A) Any employer may apply to the Secretary for a temporary order granting a variance from a standard or any provision thereof promulgated under this section. Such temporary order shall be granted only if the employer files an application which meets the requirements of clause (B) and establishes that (i) he is unable to comply with a standard by its effective date because of unavailability of professional or technical personnel or of materials and equipment needed to come into compliance with the standard or because necessary construction or alteration of facilities cannot be completed by the effective date, (ii) he is taking all available steps to safeguard his employees against the hazards covered by the standard, and (iii) he has an effective program for coming into compliance with the standard as quickly as practicable. Any temporary order issued under this paragraph shall prescribe the practices, means, methods, operations, and processes which the employer must adopt and use while the order is in effect and state in detail his program for coming into compliance with the standard. Such a temporary order may be granted only after notice to employees and an opportunity for a hearing: Provided that the Secretary may issue one interim order to be effective until a decision is made on the basis of the hearing. No temporary order may be in effect for longer than the period needed by the employer to achieve compliance with the standard or one year, whichever is shorter, except that such an order may be renewed not more that twice (I) so long as the requirements of this paragraph are met and (II) if an application for renewal is filed at least 90 days prior to the expiration date of the order. No interim renewal of an order may remain in effect for longer than 180 days.

(B) An application for temporary order under this paragraph (6) shall contain:

(i) A specification of the standard or portion thereof from which the employer seeks a variance,

(ii) A representation by the employer, supported by representations from qualified persons having firsthand knowledge of the facts represented, that he is unable to comply with the standard or portion thereof and a detailed statement of the reasons therefor,

(iii) A statement of the steps he has taken and will take (with specific dates) to protect employees against the hazard covered by the standard,

(iv) A statement of when he expects to be able to comply with the standard and what steps he has taken and what steps he will take (with dates specified) to come into compliance with the standard, and

(v) A certification that he has informed his employees of the application by giving a copy thereof to their authorized representative, posting a statement giving a summary of the application and specifying where a copy may be examined at the place or places where notices to employees are normally posted, and by other appropriate means. A description of how employees have been informed shall be contained in the certification. The information to employees shall also inform them of their right to petition the Secretary for a hearing.

(C) The Secretary is authorized to grant a variance from any standard or portion thereof whenever he determines, or the Secretary of Health, Education, and Welfare certifies, that such variance is necessary to permit an employer to participate in an experiment approved by him or the Secretary of Health, Education, and Welfare designed to demonstrate or validate new and improved techniques to safeguard the health or safety of workers.

(7) Any standard promulgated under this subsection shall prescribe the use of labels or other appropriate forms of warning as are necessary to insure that employees are apprised of all hazards to which they are exposed, relevant symptoms and appropriate emergency treatment, and proper conditions and precautions of safe use or exposure. Where appropriate, such standard shall also prescribe suitable protective equipment and control or technological procedures to be used in connection with such hazards and shall provide for monitoring or measuring employee exposure at such locations and intervals, and in such manner as may be necessary for the protection of employees. In addition, where appropriate, any such standard shall prescribe the type and frequency of medical examinations or other tests which shall be made available, by the employer or at his cost, to employees exposed to such hazards in order to most effectively determine whether the health of such employees is adversely affected by such exposure. In the event such medical examinations are in the nature of research, as determined by the Secretary of Health, Education, and Welfare, such examinations may be furnished at the expense of the Secretary of Health, Education, and Welfare. The results of such examinations or tests shall be furnished only to the Secretary or the Secretary of Health, Education, and Welfare, and, at the request of the employee, to his physician. The Secretary, in consultation with the Secretary of Health, Education, and Welfare, may by rule promulgated pursuant to section 553 of title 5, United States Code, make appropriate modifications in the foregoing requirements relating to the use of labels or other forms of warning, monitoring or measuring, and medical examinations, as may be warranted by experience, information, or medical or technological developments acquired subsequent to the promulgation of the relevant standard.

(8) Whenever a rule promulgated by the Secretary differs substantially from an existing national consensus standard, the Secretary shall, at the same time, publish in the Federal Register a statement of the reasons why the rule as adopted will better effectuate the purposes of this Act than the national consensus standard.

(c)(1) The Secretary shall provide, without regard to the requirements of chapter 5, title 5, Unites States Code, for an emergency temporary standard to take immediate effect upon publication in the Federal Register if he determines (A) that employees are exposed to grave danger from exposure to substances or agents determined to be toxic or physically harmful or from new hazards, and (B) that such emergency standard is necessary to protect employees from such danger.

(2) Such standard shall be effective until superseded by a standard promulgated in accordance with the procedures prescribed in paragraph (3) of this subsection.

(3) Upon publication of such standard in the Federal Register the Secretary shall commence a proceeding in accordance with section 6(b) of this Act, and the standard as published shall also serve as a proposed rule for the proceeding. The Secretary shall promulgate a standard under this paragraph no later than six months after publication of the emergency standard as provided in paragraph (2) of this subsection.

(d) Any affected employer may apply to the Secretary for a rule or order for a variance from a standard promulgated under this section. Affected employees shall be given notice of each such application and an opportunity to participate in a hearing. The Secretary shall issue such rule or order if he determines on the record, after opportunity for an inspection where appropriate and a hearing, that the proponent of

the variance has demonstrated by a preponderance of the evidence that the conditions, practices, means, methods, operations, or processes used or proposed to be used by an employer will provide employment and places of employment to his employees which are as safe and healthful as those which would prevail if he complied with the standard. The rule or order so issued shall prescribe the conditions the employer must maintain, and the practices, means, methods, operations, and processes which he must adopt and utilize to the extent they differ from the standard in question. Such a rule or order may be modified or revoked upon application by an employer, employees, or by the Secretary on his own motion, in the manner prescribed for its issuance under this subsection at any time after six months from its issuance.
(e) Whenever the Secretary promulgates any standard, makes any rule, order, or decision, grants any exemption or extension of time, or compromises, mitigates, or settles any penalty assessed under this Act, he shall include a statement of the reasons for such action, which shall be published in the Federal Register.
(f) Any person who may be adversely affected by a standard issued under this section may at any time prior to the sixtieth day after such standard is promulgated file a petition challenging the validity of such standard with the United States court of appeals for the circuit wherein such person resides or has his principal place of business, for a judicial review of such standard. A copy of the petition shall be forthwith transmitted by the clerk of the court to the Secretary. The filing of such petition shall not, unless otherwise ordered by the court, operate as a stay of the standard. The determinations of the Secretary shall be conclusive if supported by substantial evidence in the record considered as a whole.
(g) In determining the priority for establishing standards under this section, the Secretary shall give due regard to the urgency of the need for mandatory safety and health standards for particular industries, trades, crafts, occupations, businesses, workplaces or work environments. The Secretary shall also give due regard to the recommendations of the Secretary of Health, Education, and Welfare regarding the need for mandatory standards in determining the priority for establishing such standards.

SECTION 7—ADVISORY COMMITTEES; ADMINISTRATION

(a)(1) There is hereby established a National Advisory Committee on Occupational Safety and Health consisting of twelve members appointed by the Secretary, four of whom are to be designated by the Secretary of Health, Education, and Welfare, without regard to the provisions of title 5, United States Code, governing appointments in the competitive service, and composed of representatives of management, labor, occupational safety and occupational health professions, and of the public. The Secretary shall designate one of the public members as Chairman. The members shall be selected upon the basis of their experience and competence in the field of occupational safety and health.

(2) The Committee shall advise, consult with, and make recommendations to the Secretary and the Secretary of Health, Education, and Welfare on matters relating to the administration of the Act. The Committee shall hold no fewer than two meetings during each calendar year. All meetings of the Committee shall be open to the public and a transcript shall be kept and made available for public inspection.

(3) The members of the Committee shall be compensated in accordance with the provisions of section 3109 of title 5, United States Code.

(4) The Secretary shall furnish to the Committee an executive secretary and such secretarial, clerical, and other services as are deemed necessary to the conduct of its business.

(b) An advisory committee may be appointed by the Secretary to assist him in his standard setting functions under section 6 of this Act. Each such committee shall consist of not more than fifteen members and shall include as a member one of more designees of the Secretary of Health, Education, and Welfare, and shall include among its members an equal number of persons qualified by experience and affiliation to present the viewpoint of the employers involved, and of persons similarly qualified to present the viewpoint of the workers involved, as well as one or more representatives of health and safety agencies of the States. An advisory committee may also include such other persons as the Secretary may appoint who are qualified by knowledge and experience to make a useful contribution to the work of such committee, including one or more representatives of professional organizations of technicians or professionals specializing in occupational safety or health, and one or more representatives of nationally recognized standards-producing organizations, but the number of persons so appointed to any such advisory committee shall not exceed the number appointed to such committee as representatives of Federal and State agencies. Persons appointed to advisory committees from private life shall be compensated in the same manner as consultants or experts under section 3109 of title 5, United States Code. The Secretary shall pay to any State which is the employer of a member of such a committee who is a representative of the health or safety agency of that State, reimbursement sufficient to cover the actual cost to the State resulting from such representative's membership on such committee. Any meeting of such committee shall be open to the public and an accurate record shall be kept and made available to the public. No member of such committee (other than representatives of employers and employees) shall have an economic interest in any proposed rule.

(c) In carrying out his responsibilities under this Act, the Secretary is authorized to:

(1) Use, with the consent of any Federal agency, the services, facilities, and personnel of such agency, with or without reimbursement, and with the consent of any State or political subdivision thereof, accept and use the services, facilities, and personnel of any agency of such State of subdivision with reimbursement; and

(2) Employ experts and consultants or organizations thereof as authorized by section 3109 of title 5, United States Code, except that contracts for such employment may be renewed annually; compensate individuals so employed at rates not in excess of the rate specified at the time of service for grade GS-18 under section 5332 of title 5, United States Code, including travel time, and allow them while away from their homes or regular places of business, travel expenses (including per diem in lieu of subsistence) as authorized by section 5703 of title 5, United States Code, for persons in the Government service employed intermittently, while so employed.

SECTION 8—INSPECTIONS, INVESTIGATIONS, AND RECORDKEEPING

(a) In order to carry out the purposes of this Act, the Secretary, upon presenting appropriate credentials to the owner, operator, or agent in charge, is authorized:

(1) To enter without delay and at reasonable times any factory, plant, establishment, construction site, or other area, workplace or environment where work is performed by an employee of an employer; and

(2) To inspect and investigate during regular working hours and at other reasonable times, and within reasonable limits and in a reasonable manner, any such place of employment and all pertinent conditions, structures, machines, apparatus, devices, equipment, and materials therein, and to question privately any such employer, owner, operator, agent or employee.

(b) In making his inspections and investigations under this Act the Secretary may require the attendance and testimony of witnesses and the production of evidence under oath. Witnesses shall be paid the same fees and mileage that are paid witnesses in the courts of the United States. In case of a contumacy, failure, or refusal of any person to obey such an order, any district court of the United States or the United States courts of any territory or possession, within the jurisdiction of which such person is found, or resides or transacts business, upon the application by the Secretary, shall have jurisdiction to issue to such person an order requiring such person to appear to produce evidence if, as, and when so ordered, and to give testimony relating to the matter under investigation or in question, and any failure to obey such order of the court may be punished by said court as a contempt thereof.

(c)(1) Each employer shall make, keep and preserve, and make available to the Secretary or the Secretary of Health, Education, and Welfare, such records regarding his activities relating to this Act as the Secretary, in cooperation with the Secretary of Health, Education, and Welfare, may prescribe by regulation as necessary or appropriate for the enforcement of this Act or for developing information regarding the causes and prevention of occupational accidents and illnesses. In order to carry out the provisions of this paragraph such regulations may include provisions requiring employers to conduct periodic inspections. The Secretary shall also issue regulations requiring that employers, through posting of notices or other appropriate means, keep their employees informed of their protections and obligations under this Act, including the provisions of applicable standards.

(2) The Secretary, in cooperation with the Secretary of Health, Education and Welfare, shall issue regulations requiring employers to maintain accurate records of, and to make periodic reports on, work-related deaths, injuries and illnesses other than minor injuries requiring only first aid treatment and which do not involve medical treatment, loss of consciousness, restriction of work or motion, or transfer to another job.

(3) The Secretary, in cooperation with the Secretary of Health, Education, and Welfare, shall issue regulations requiring employers to maintain accurate records of employee exposures to potentially toxic materials or harmful physical agents which are required to be monitored or measured under section 6. Such regulations shall provide employees or their representatives with an opportunity to observe such monitoring or measuring, and to have access to the records thereof. Such regulations shall also make appropriate provision for each employee or former employee to have access to such records as will

indicate his own exposure to toxic materials or harmful physical agents. Each employer shall promptly notify any employee who has been or is being exposed to toxic materials or harmful physical agents in concentrations or at levels which exceed those prescribed by an applicable occupational safety and health standard promulgated under section 6, and shall inform any employee who is being thus exposed of the corrective action being taken.

(d) Any information obtained by the Secretary, the Secretary of Health, Education and Welfare, or a State agency under this Act shall be obtained with a minimum burden upon employers, especially those operating small businesses. Unnecessary duplication of efforts in obtaining information shall be reduced to the maximum extent feasible.

(e) Subject to regulations issued by the Secretary, a representative of the employer and a representative authorized by his employees shall be given an opportunity to accompany the Secretary or his authorized representative during the physical inspection of any workplace under subsection (a) for the purpose of aiding such inspection. Where there is no authorized employee representative, the Secretary or his authorized representative shall consult with a reasonable number of employees concerning matters of health and safety in the workplace.

(f)(1) Any employees or representative of employees who believe that a violation of a safety or health standard exists that threatens physical harm, or that an imminent danger exists, may request an inspection by giving notice to the Secretary or his authorized representative of such violation or danger. Any such notice shall be reduced to writing, shall set forth with reasonable particularity the grounds for the notice, and shall be signed by the employees or representative of employees, and a copy shall be provided the employer or his agent no later than at the time of inspection, except that, upon the request of the person giving such notice, his name and the names of individual employees referred to therein shall not appear in such copy or on any record published, released, or made available pursuant to subsection (g) of this section. If upon receipt of such notification the Secretary determines there are reasonable grounds to believe that such violation or danger exists, he shall make a special inspection in accordance with the provisions of this section as soon as practicable, to determine if such violation or danger exists. If the Secretary determines there are no reasonable grounds to believe that a violation or danger exists he shall notify the employees or representative of the employees in writing of such determination.

(2) Prior to or during any inspection of a workplace, any employees or representative of employees employed in such workplace may notify the Secretary or any representative of the Secretary responsible for conducting the inspection, in writing, of any violation of this Act which they have reason to believe exists in such workplace. The Secretary shall, by regulation, establish procedures for informal review of any refusal by a representative of the Secretary to issue a citation with respect to any such alleged violation and shall furnish the employees or representative of employees requesting such review a written statement of the reasons for the Secretary's final disposition of the case.

(g)(1) The Secretary and Secretary of Health, Education, and Welfare are authorized to compile, analyze, and publish, either in summary or detailed form, all reports or information obtained under this section.

(2) The Secretary and the Secretary of Health, Education, and Welfare shall each prescribe such rules and regulations as he may deem necessary to carry out their responsibilities under this Act, including rules and regulations dealing with the inspection of an employer's establishment.

SECTION 9—CITATIONS

(a) If, upon inspection or investigation, the Secretary or his authorized representative believes that an employer has violated a requirement of section 5 of this Act, of any standard, rule or order promulgated pursuant to section 6 of this Act, or of any regulations prescribed pursuant to this Act, he shall with reasonable promptness issue a citation to the employer. Each citation shall be in writing and shall describe with particularity the nature of the violation, including a reference to the provision of the Act, standard, rule, regulation, or order alleged to have been violated. In addition, the citation shall fix a reasonable time for the abatement of the violation. The Secretary may prescribe procedures for the issuance of a notice in lieu of a citation with respect to de minimis violations which have no direct or immediate relationship to safety or health.

(b) Each citation issued under this section, or a copy or copies thereof, shall be prominently posted, as prescribed in regulations issued by the Secretary, at or near each place a violation referred to in the citation occurred.

(c) No citation may be issued under this section after the expiration of six months following the occurrence of any violation.

SECTION 10—PROCEDURE FOR ENFORCEMENT

(a) If, after an inspection or investigation, the Secretary issues a citation under section 9(a), he shall, within a reasonable time after the termination of such inspection or investigation, notify the employer by certified mail of the penalty, if any, proposed to be assessed under section 17 and that the employer has fifteen working days within which to notify the Secretary that he wishes to contest the citation or proposed assessment of penalty. If, within fifteen working days from the receipt of the notice issued by the Secretary the employer fails to notify the Secretary that he intends to contest the citation or proposed assessment of penalty, and no notice is filed by any employees or representative of employees under subsection (c) within such time, the citation and the assessment, as proposed, shall be deemed a final order of the Commission and not subject to review by any court or agency.
(b) If the Secretary has reason to believe that an employer has failed to correct a violation for which a citation has been issued within the period permitted for its correction (which period shall not begin to run until the entry of a final order by the Commission in the case of any review proceedings under this section initiated by the employer in good faith and not solely for delay or avoidance of penalties), the Secretary shall notify the employer by certified mail of such failure and of the penalty proposed to be assessed under section 17 by reason of such failure, and that the employer has fifteen working days within which to notify the Secretary that he wishes to contest the Secretary's notification or the proposed assessment of penalty. If, within fifteen working days from the receipt of notification issued by the Secretary, the employer fails to notify the Secretary that he intends to contest the notification or proposed assessment of penalty, the notification and assessment, as proposed, shall be deemed a final order of the Commission and not subject to review by any court or agency.
(c) If an employer notifies the Secretary that he intends to contest a citation issued under section 9(a) or notification issued under subsection (a) or (b) of this section, or if, within fifteen working days of the issuance of a citation under section 9(a), any employee or representative of employees files a notice with the Secretary alleging that the period of time fixed in the citation for the abatement of the violation is unreasonable, the Secretary shall immediately advise the Commission of such notification, and the Commission shall afford an opportunity for a hearing (in accordance with section 554 of title 5, United States Code, but without regard to subsection (a)(3) of such section). The Commission shall thereafter issue an order, based on findings of fact, affirming, modifying, or vacating the Secretary's citation or proposed penalty, or directing other appropriate relief, and such order shall become final thirty days after its issuance. Upon a showing by an employer of a good faith effort to comply with the abatement requirements of a citation, and that abatement has not been completed because of factors beyond his reasonable control, the Secretary, after an opportunity for a hearing as provided in this subsection, shall issue an order affirming or modifying the abatement requirements in such citation. The rules of procedure prescribed by the Commission shall provide affected employees or representatives of affected employees an opportunity to participate as parties to hearings under this subsection.

SECTION 11—JUDICIAL REVIEW

(a) Any person adversely affected or aggrieved by an order of the Commission issued under subsection (c) of section 10 may obtain a review of such order in any United States court of appeals for the circuit in which the violation is alleged to have occurred or where the employer has its principal office, or in the Court of Appeals for the District of Columbia Circuit, by filing in such court within sixty days following the issuance of such order a written petition praying that the order be modified or set aside. A copy of such petition shall be forthwith transmitted by the clerk of the court to the Commission and to the other parties, and thereupon the Commission shall file in the court the record in the proceeding as provided in section 2112 of title 28, United States Code. Upon such filing, the court shall have jurisdiction of the proceeding and of the question determined therein, and shall have power to grant such temporary relief or restraining order as it deems just and proper, and to make and enter upon the pleadings, testimony, and proceedings set forth in such record a decree affirming, modifying, or setting aside in whole or in part, the order of the Commission and enforcing the same to the extent that such order is affirmed or modified. The commencement of proceedings under this subsection shall not, unless ordered by the court, operate as a stay of the order of the Commission. No objection that has not been urged before the Commission shall be considered by the court, unless the failure or neglect to urge such objection shall be excused because of extraordinary circumstances. The findings of the Commission with respect to questions of fact, if supported by substantial evidence on the record considered as a whole, shall be conclusive. If any party shall apply to the court for leave to adduce additional evidence and shall show to the satisfaction of the court that such additional evidence is material and that there were reasonable grounds for the failure to adduce such evidence in the hearing before the Commission, the court may order such additional evidence to be taken before the Commission and to be made a part of the record. The Commission may modify its findings as to the facts, or make new findings, by reason of additional evidence so taken and filed, and it shall file such modified or new findings, which findings with respect to questions of fact, if supported by substantial evidence on the record considered as a whole, shall be conclusive, and its recommendations, if any, for the modification or setting aside of its original order. Upon the filing of the record with it, the jurisdiction of the court shall be exclusive and its judgment and decree shall be final, except that the same shall be subject to review by the Supreme Court of the United States, as provided in section 1254 of title 28, United States Code. Petitions filed under this subsection shall be heard expeditiously.

(b) The Secretary may also obtain review or enforcement of any final order of the Commission by filing a petition for such relief in the United States court of appeals for the circuit in which the alleged violation occurred or in which the employer has its principal office, and the provisions of subsection (a) shall govern such proceedings to the extent applicable. If no petition for review, as provided in subsection (a), is filed within sixty days after service of the Commission's order, the Commission's findings of fact and order shall be conclusive in connection with any petition for enforcement which is filed by the Secretary after the expiration of such sixty-day period. In any such case, as well as in the case of a non-contested citation or notification by the Secretary which has become a final order of the Commission under subsection (a) or (b) of section 10, the clerk of the court, unless otherwise ordered by the court, shall forthwith enter a decree enforcing the order and shall transmit a copy of such decree to the Secretary and the employer named in the petition. In any contempt proceeding brought to enforce a decree of a court of appeals entered pursuant to this subsection or subsection (a), the court of appeals may assess the penalties provided in section 17, in addition to invoking any other available remedies.

(c)(1) No person shall discharge or in any manner discriminate against any employee because such employee has filed any complaint or instituted or caused to be instituted any proceeding under or related to this Act or has testified or is about to testify in any such proceeding or because of the exercise by such employee on behalf of himself or others of any right afforded by this Act.

(2) Any employee who believes that he has been discharged or otherwise discriminated against by any person in violation of this subsection may, within thirty days after such violation occurs, file a complaint with the Secretary alleging such discrimination. Upon receipt of such complaint, the Secretary shall cause such investigation to be made as he deems appropriate. If upon such investigation, the Secretary determines that the provisions of this subsection have been violated, he shall bring an action in any appropriate United States district court against such person. In any such action the United States district courts shall have jurisdiction, for cause shown to restrain violations of paragraph (1) of this subsection and order all appropriate relief including rehiring or reinstatement of the employee to his former position with back pay.

(3) Within 90 days of the receipt of a complaint filed under this subsection the Secretary shall notify the complainant of his determination under paragraph 2 of this subsection.

SECTION 12—THE OCCUPATIONAL SAFETY AND HEALTH REVIEW COMMISSION

(a) The Occupational Safety and Health Review Commission is hereby established. The Commission shall be composed of three members who shall be appointed by the President, by and with the advice and consent of the Senate, from among persons who by reason of training, education, or experience are qualified to carry out the functions of the Commission under this Act. The President shall designate one of the members of the Commission to serve as Chairman.

(b) The terms of members of the Commission shall be six years except that (1) the members of the Commission first taking office shall serve, as designated by the President at the time of appointment, one for a term of two years, one for a term of four years, and one for a term of six years, and (2) a vacancy caused by the death, resignation, or removal of a member prior to the expiration of the term for which he was appointed shall be filled only for the remainder of such unexpired term. A member of the Commission may be removed by the President for inefficiency, neglect of duty, or malfeasance in office.

(c)(1) Section 5314 of title 5, United States Code, is amended by adding at the end thereof the following new paragraph:

"(57) Chairman, Occupational Safety and Health Review Commission."
(2) Section 5315 of title 5, United States Code, is amended by adding at the end thereof the following new paragraph:
"(94) Members, Occupational Safety and Health Review Commission."
(d) The principal office of the Commission shall be in the District of Columbia. Whenever the Commission deems that the convenience of the public or of the parties may be promoted, or delay or expense may be minimized, it may hold hearings or conduct other proceedings at any other place.
(e) The Chairman shall be responsible on behalf of the Commission for the administrative operations of the Commission and shall appoint such hearing examiners and other employees as he deems necessary to assist in the performance of the Commission's functions and to fix their compensation in accordance with the provisions of chapter 51 and subchapter III of chapter 53 of title 5, United States Code, relating to classification and General Schedule pay rates: Provided, That assignment, removal and compensation of hearing examiners shall be in accordance with sections 3105, 3344, 5362, and 7521 of title 5, United States Code.
(f) For the purpose of carrying out its functions under this Act, two members of the Commission shall constitute a quorum and official action can be taken only on the affirmative vote of at least two members.
(g) Every official act of the Commission shall be entered of record, and its hearings and records shall be open to the public. The Commission is authorized to make such rules as are necessary for the orderly transaction of its proceedings. Unless the Commission has adopted a different rule, its proceedings shall be in accordance with the Federal Rules of Civil Procedure.
(h) The Commission may order testimony to be taken by deposition in any proceedings pending before it at any state of such proceeding. Any person may be compelled to appear and depose, and to produce books, papers, or documents, in the same manner as witnesses may be compelled to appear and testify and produce like documentary evidence before the Commission. Witnesses whose depositions are taken under this subsection, and the persons taking such depositions, shall be entitled to the same fees as are paid for like services in the courts of the United States.
(i) For the purpose of any proceeding before the Commission, the provisions of section 11 of the National Labor Relations Act (29 U.S.C. 161) are hereby made applicable to the jurisdiction and powers of the Commission.
(j) A hearing examiner appointed by the Commission shall hear, and make a determination upon, any proceeding instituted before the Commission and any motion in connection therewith, assigned to such hearing examiner by the Chairman of the Commission, and shall make a report of any such determination which constitutes his final disposition of the proceedings. The report of the hearing examiner shall become the final order of the Commission within thirty days after such report by the hearing examiner, unless within such period any Commission member has directed that such report shall be reviewed by the Commission.
(k) Except as otherwise provided in this Act, the hearing examiners shall be subject to the laws governing employees in the classified civil service, except that appointments shall be made without regard to section 5108 of title 5, United States Code. Each hearing examiner shall receive compensation at a rate not less than that prescribed for GS-16 under section 5332 of title 5, United States Code.

SECTION 13—PROCEDURES TO COUNTERACT IMMINENT DANGERS

(a) The United States district courts shall have jurisdiction, upon petition of the Secretary, to restrain any conditions or practices in any place of employment which are such that a danger exists which could reasonably be expected to cause death or serious physical harm immediately or before the imminence of such danger can be eliminated through the enforcement procedures otherwise provided by this Act. Any order issued under this section may require such steps to be taken as may be necessary to avoid, correct, or remove such imminent danger and prohibit the employment or presence of any individual in locations or under conditions where such imminent danger exists, except individuals whose presence is necessary to avoid, correct, or remove such imminent danger or to maintain the capacity of a continuous process operation to resume normal operations without a complete cessation of operations, or where a cessation of operations is necessary, to permit such to be accomplished in a safe and orderly manner.
(b) Upon the filing of any such petition the district court shall have jurisdiction to grant such injunctive relief or temporary restraining order pending the outcome of an enforcement proceeding pursuant to this Act. The proceeding shall be as provided by Rule 65 of the Federal Rules, Civil Procedure, except that no temporary restraining order issued without notice shall be effective for a period longer than five days.

(c) Whenever and as soon as an inspector concludes that conditions or practices described in subsection (a) exist in any place of employment, he shall inform the affected employees and employers of the danger and that he is recommending to the Secretary that relief be sought.
(d) If the Secretary arbitrarily or capriciously fails to seek relief under this section, any employee who may be injured by reason of such failure, or the representative of such employees, might bring an action against the Secretary in the United States district court for the district in which the imminent danger is alleged to exist or the employer has its principal office, or for the District of Columbia, for a writ of mandamus to compel the Secretary to seek such an order and for such further relief as may be appropriate.

SECTION 14—REPRESENTATION IN CIVIL LITIGATION

Except as provided in section 518(a) of title 28, United States Code, relating to litigation before the Supreme Court, the Solicitor of Labor may appear for and represent the Secretary in any civil litigation brought under this Act but all such litigation shall be subject to the direction and control of the Attorney General.

SECTION 15—CONFIDENTIALITY OF TRADE SECRETS

All information reported to or otherwise obtained by the Secretary or his representative in connection with any inspection or proceeding under this Act which contains or which might reveal a trade secret referred to in section 1905 of title 18 of the United States Code shall be considered confidential for the purpose of that section, except that such information may be disclosed to other officers or employees concerned with carrying out this Act or when relevant in any proceeding under this Act. In any such proceeding the Secretary, the Commission, or the court shall issue such orders as may be appropriate to protect the confidentiality of trade secrets.

SECTION 16—VARIATIONS, TOLERANCES, AND EXEMPTIONS

The Secretary, on the record, after notice and opportunity for a hearing may provide such reasonable limitations and may make such rules and regulations allowing reasonable variations, tolerances, and exemptions to and from any or all provisions of this Act as he may find necessary and proper to avoid serious impairment of the national defense. Such action shall not be in effect for more than six months without notification to affected employees and an opportunity being afforded for a hearing.

SECTION 17—PENALTIES

(a) Any employer who willfully or repeatedly violates the requirements of section 5 of this Act, any standard, rule, or order promulgated pursuant to section 6 of this Act, or regulations prescribed pursuant to this Act, may be assessed a civil penalty of not more than $70,000 for each violation, but not less than $5,000 for each willful violation.
(b) Any employer who has received a citation for a serious violation of the requirements of section 5 of this Act, of any standard, rule, or order promulgated pursuant to section 6 of this Act, or of any regulations prescribed pursuant to this Act, shall be assessed a civil penalty of up to $7,000 for each such violation.
(c) Any employer who has received a citation for a violation of the requirements of section 5 of this Act, of any standard, rule, or order promulgated pursuant to section 6 of this Act, or of regulations prescribed pursuant to this Act, and such violation is specifically determined not to be of a serious nature, may be assessed a civil penalty of up to $7,000 for each violation.
(d) Any employer who fails to correct a violation for which a citation has been issued under section 9(a) within the period permitted for its correction (which period shall not begin to run until the date of the final order of the Commission in the case of any review proceeding under section 10 initiated by the employer in good faith and not solely for delay or avoidance of penalties), may be assessed a civil penalty of not more than $7,000 for each day during which such failure or violation continues.
(e) Any employer who willfully violates any standard, rule, or order promulgated pursuant to section 6 of this Act, or of any regulations prescribed pursuant to this Act, and that violation caused death to any employee, shall, upon conviction, be punished by a fine of not more than $10,000 or by imprisonment for not more than six months, or by both; except that if the conviction is for a violation committed after a

first conviction of such person, punishment shall be by a fine of not more than $20,000 or by imprisonment for not more than one year, or by both.
(f) Any person who gives advance notice of any inspection to be conducted under this Act, without authority from the Secretary or his designees, shall, upon conviction, be punished by a fine of not more than $1,000 or by imprisonment for not more than six months, or by both.
(g) Whoever knowingly makes any false statement, representation, or certification in any application, record, report, plan, or other document filed or required to be maintained pursuant to this Act shall, upon conviction, be punished by a fine of not more than $10,000, or by
imprisonment for not more than six months, or by both.
(h)(1) Section 1114 of title 18, United States Code, is hereby amended by striking out "designated by the Secretary of Health, Education, and Welfare to conduct investigations, or inspections under the Federal Food, Drug, and Cosmetic Act" and inserting in lieu thereof "or of the Department of Labor assigned to perform investigative, inspection, or law enforcement functions".

(2) Notwithstanding the provisions of sections 1111 and 1114 of title 18, United States Code, whoever, in violation of the provisions of section 1114 of such title, kills a person while engaged in or on account of the performance of investigative, inspection, or law enforcement functions added to such section 1114 by paragraph (1) of this subsection, and who would otherwise be subject to the penalty provisions of such section 1111, shall be punished by imprisonment for any term of years or for life.
(i) Any employer who violates any of the posting requirements, as prescribed under the provisions of this Act, shall be assessed a civil penalty of up to $7,000 for each violation.
(j) The Commission shall have authority to assess all civil penalties provided in this section, giving due consideration to the appropriateness of the penalty with respect to the size of the business of the employer being charged, the gravity of the violation, the good faith of the employer, and the history of previous violations.
(k) For purposes of this section, a serious violation shall be deemed to exist in a place of employment if there is a substantial probability that death or serious physical harm could result from a condition which exists, or from one or more practices, means, methods, operations, or processes which have been adopted or are in use, in such place of employment unless the employer did not, and could not with the exercise of reasonable diligence, know of the presence of the violation.
(l) Civil penalties owed under this Act shall be paid to the Secretary for deposit into the Treasury of the United States and shall accrue to the United States and may be recovered in a civil action in the name of the United States brought in the United States district court for the district where the violation is alleged to have occurred or where the employer has its principal office.

SECTION 18—STATE JURISDICTION AND STATE PLANS

(a) Nothing in this Act shall prevent any State agency or court from asserting jurisdiction under State law over any occupational safety or health issue with respect to which no standard is in effect under section 6.
(b) Any State which, at any time, desires to assume responsibility for development and enforcement therein of occupational safety and health standards relating to any occupational safety or health issue with respect to which a Federal standard has been promulgated under section 6 shall submit a State plan for the development of such standards and their enforcement.
(c) The Secretary shall approve the plan submitted by a State under subsection (b), or any modification thereof, if such plan in his judgment:

(1) designates a State agency or agencies as the agency or agencies responsible for administering the plan throughout the State,

(2) provides for the development and enforcement of safety and health standards relating to one or more safety or health issues, which standards (and the enforcement of which standards) are or will be at least as effective in providing safe and healthful employment and places of employment as the standards promulgated under section 6 which relate to the same issues, and which standards, when applicable to products which are distributed or used in interstate commerce, are required by compelling local conditions and do not unduly burden interstate commerce,

(3) provides for a right of entry and inspection of all workplaces subject to the Act which is at least as effective as that provided in section 8, and includes a prohibition on advance notice of inspections,

(4) contains satisfactory assurances that such agency or agencies have or will have the legal authority and qualified personnel necessary for the enforcement of such standards,

(5) gives satisfactory assurances that such State will devote adequate funds to the administration and enforcement of such standards,

(6) contains satisfactory assurances that such State will, to the extent permitted by its law, establish and maintain an effective and comprehensive occupational safety and health program applicable to all employees of public agencies of the State and its political subdivisions, which program is as effective as the standards contained in an approved plan,

(7) requires employers in the State to make reports to the Secretary in the same manner and to the same extent as if the plan were not in effect, and

(8) provides that the State agency will make such reports to the Secretary in such form and containing such information, as the Secretary shall from time to time require.

(d) If the Secretary rejects a plan submitted under subsection (b), he shall afford the State submitting the plan due notice and opportunity for a hearing before so doing.

(e) After the Secretary approves a State plan submitted under subsection (b), he may, but shall not be required to, exercise his authority under sections 8, 9, 10, 13, and 17 with respect to comparable standards promulgated under section 6, for the period specified in the next sentence. The Secretary may exercise the authority referred to above until he determines, on the basis of actual operations under the State plan, that the criteria set forth in subsection (c) are being applied, but he shall not make such determination for at least three years after the plan's approval under subsection (c). Upon making the determination referred to in the preceding sentence, the provisions of sections 5(a)(2), 8 (except for the purpose of carrying out subsection (f) of this section), 9, 10, 13, and 17, and standards promulgated under section 6 of this Act, shall not apply with respect to any occupational safety or health issues covered under the plan, but the Secretary may retain jurisdiction under the above provisions in any proceeding commenced under section 9 or 10 before the date of determination.

(f) The Secretary shall, on the basis of reports submitted by the State agency and his own inspections make a continuing evaluation of the manner in which each State having a plan approved under this section is carrying out such plan. Whenever the Secretary finds, after affording due notice and opportunity for a hearing, that in the administration of the State plan there is a failure to comply substantially with any provision of the State plan (or any assurance contained therein), he shall notify the State agency of his withdrawal of approval of such plan and upon receipt of such notice such plan shall cease to be in effect, but the State may retain jurisdiction in any case commenced before the withdrawal of the plan in order to enforce standards under the plan whenever the issues involved do not relate to the reasons for the withdrawal of the plan.

(g) The State may obtain a review of a decision of the Secretary withdrawing approval of or rejecting its plan by the United States court of appeals for the circuit in which the State is located by filing in such court within thirty days following receipt of notice of such decision a petition to modify or set aside in whole or in part the action of the Secretary. A copy of such petition shall forthwith be served upon the Secretary, and thereupon the Secretary shall certify and file in the court the record upon which the decision complained of was issued as provided in section 2112 of title 28, United States Code. Unless the court finds that the Secretary's decision in rejecting a proposed State plan or withdrawing his approval of such a plan is not supported by substantial evidence the court shall affirm the Secretary's decision. The judgment of the court shall be subject to review by the Supreme Court of the United States upon certiorari or certification as provided in section 1254 of title 28, United States Code.

(h) The Secretary may enter into an agreement with a State under which the State will be permitted to continue to enforce one or more occupational health and safety standards in effect in such State until final action is taken by the Secretary with respect to a plan submitted by a State under subsection (b) of this section, or two years from the date of enactment of this Act, whichever is earlier.

SECTION 19—FEDERAL AGENCY SAFETY PROGRAMS AND RESPONSIBILITIES

(a) It shall be the responsibility of the head of each Federal agency to establish and maintain an effective and comprehensive occupational safety and health program which is consistent with the standards promulgated under section 6. The head of each agency shall (after consultation with representatives of the employees thereof):

(1) provide safe and healthful places and conditions of employment, consistent with the standards set under section 6;

(2) acquire, maintain, and require the use of safety equipment, personal protective equipment, and devices reasonably necessary to protect employees;

(3) keep adequate records of all occupational accidents and illnesses for proper evaluation and necessary corrective action;

(4) consult with the Secretary with regard to the adequacy as to form and content of records kept pursuant to subsection (a)(3) of this section; and

(5) make an annual report to the Secretary with respect to occupational accidents and injuries and the agency's program under this section. Such report shall include any report submitted under section 7902(e)(2) of title 5, United States Code.

(b) The Secretary shall report to the President a summary or digest of reports submitted to him under subsection (a)(5) of this section, together with his evaluations of and recommendations derived from such reports. The President shall transmit annually to the Senate and the House of Representatives a report of the activities of Federal agencies under this section.

(c) Section 7902(c)(1) of title 5, United States Code, is amended by inserting after "agencies" the following: "and of labor organizations representing employees".

(d) The Secretary shall have access to records and reports kept and filed by Federal agencies pursuant to subsections (a)(3) and (5) of this section unless those records and reports are specifically required by Executive order to be kept secret in the interest of the national defense or foreign policy, in which case the Secretary shall have access to such information as will not jeopardize national defense or foreign policy.

SECTION 20—RESEARCH AND RELATED ACTIVITIES

(a)(1) The Secretary of Health, Education, and Welfare, after consultation with the Secretary and with other appropriate Federal departments or agencies, shall conduct (directly or by grants or contracts) research, experiments, and demonstrations relating to occupational safety and health, including studies of psychological factors involved, and relating to innovative methods, techniques, and approaches for dealing with occupational safety and health problems.

(2) The Secretary of Health, Education, and Welfare shall from time to time consult with the Secretary in order to develop specific plans for such research, demonstrations, and experiments as are necessary to produce criteria, including criteria identifying toxic substances, enabling the Secretary to meet his responsibility for the formulation of safety and health standards under this Act; and the Secretary of Health, Education, and Welfare, on the basis of such research, demonstrations, and experiments and any other information available to him, shall develop and publish at least annually such criteria as will effectuate the purposes of this Act.

(3) The Secretary of Health, Education, and Welfare, on the basis of such research, demonstrations, and experiments, and any other information available to him, shall develop criteria dealing with toxic materials and harmful physical agents and substances which will describe exposure levels that are safe for various periods of employment, including but not limited to the exposure levels at which no employee will suffer impaired health or functional capacities or diminished life expectancy as a result of his work experience.

(4) The Secretary of Health, Education, and Welfare shall also conduct special research, experiments, and demonstrations relating to occupational safety and health as are necessary to explore new problems, including those created by new technology in occupational safety and health, which may require ameliorative action beyond that which is otherwise provided for in the operating provisions of this Act. The Secretary of Health, Education, and Welfare shall also conduct research into the motivational and behavioral factors relating to the field of occupational safety and health.

(5) The Secretary of Health, Education, and Welfare, in order to comply with his responsibilities under paragraph (2), and in order to develop needed information regarding potentially toxic substances or harmful physical agents, may prescribe regulations requiring employers to measure, record, and make reports on the exposure of employees to substances or physical agents which the Secretary of Health, Education, and Welfare reasonably believes may endanger the health or safety of employees. The Secretary of Health, Education, and Welfare also is authorized to establish such programs of medical examinations and tests as may be necessary for determining the incidence of occupational illnesses and the susceptibility of employees to such illnesses. Nothing in this or any other provision of this Act shall be deemed to authorize or require medical examination, immunization, or treatment for those who object thereto on religious grounds, except where such is necessary for the protection of the health or safety of others. Upon the request of any employer who is required to measure and record exposure of employees to substances or physical agents as provided under this subsection, the Secretary of Health, Education, and Welfare shall furnish full financial or other assistance to such employer for the purpose of defraying

any additional expense incurred by him in carrying out the measuring and recording as provided in this subsection.

(6) The Secretary of Health, Education, and Welfare shall publish within six months of enactment of this Act and thereafter as needed but at least annually a list of all known toxic substances by generic family or other useful grouping, and the concentrations at which such toxicity is known to occur. He shall determine following a written request by any employer or authorized representative of employees, specifying with reasonable particularity the grounds on which the request is made, whether any substance normally found in the place of employment has potentially toxic effects in such concentrations as used or found; and shall submit such determination both to employers and affected employees as soon as possible. If the Secretary of Health, Education, and Welfare determines that any substance is potentially toxic at the concentrations in which it is used or found in a place of employment, and such substance is not covered by an occupational safety or health standard promulgated under section 6, the Secretary of Health, Education, and Welfare shall immediately submit such determination to the Secretary, together with all pertinent criteria.

(7) Within two years of enactment of the Act, and annually thereafter the Secretary of Health, Education, and Welfare shall conduct and publish industry-wide studies of the effect of chronic or low-level exposure to industrial materials, processes, and stresses on the potential for illness, disease, or loss of functional capacity in aging adults.

(b) The Secretary of Health, Education, and Welfare is authorized to make inspections and question employers and employees as provided in section 8 of this Act in order to carry out his functions and responsibilities under this section.

(c) The Secretary is authorized to enter into contracts, agreements, or other arrangements with appropriate public agencies or private organizations for the purpose of conducting studies relating to his responsibilities under this Act. In carrying out his responsibilities under this subsection, the Secretary shall cooperate with the Secretary of Health, Education, and Welfare in order to avoid any duplication of efforts under this section.

(d) Information obtained by the Secretary and the Secretary of Health, Education, and Welfare under this section shall be disseminated by the Secretary to employers and employees and organizations thereof.

(e) The functions of the Secretary of Health, Education, and Welfare under this Act shall, to the extent feasible, be delegated to the Director of the National Institute for Occupational Safety and Health established by section 22 of this Act.

SECTION 21—TRAINING AND EMPLOYEE EDUCATION

(a) The Secretary of Health, Education, and Welfare, after consultation with the Secretary and with other appropriate Federal departments and agencies, shall conduct, directly or by grants or contracts (1) education programs to provide an adequate supply of qualified personnel to carry out the purposes of this Act, and (2) informational programs on the importance of and proper use of adequate safety and health equipment.

(b) The Secretary is also authorized to conduct, directly or by grants or contracts, short-term training of personnel engaged in work related to his responsibilities under this Act.

(c) The Secretary, in consultation with the Secretary of Health, Education, and Welfare, shall (1) provide for the establishment and supervision of programs for the education and training of employers and employees in the recognition, avoidance, and prevention of unsafe or unhealthful working conditions in employments covered by this Act, and (2) consult with and advise employers and employees, and organizations representing employers and employees as to effective means of preventing occupational injuries and illnesses.

SECTION 22—NATIONAL INSTITUTE FOR OCCUPATIONAL SAFETY AND HEALTH

(a) It is the purpose of this section to establish a National Institute for Occupational Safety and Health in the Department of Health, Education, and Welfare in order to carry out the policy set forth in section 2 of this Act and to perform the functions of the Secretary of Health, Education, and Welfare under sections 20 and 21 of this Act.

(b) There is hereby established in the Department of Health, Education, and Welfare a National Institute for Occupational Safety and Health. The Institute shall be headed by a Director who shall be appointed by the Secretary of Health, Education, and Welfare, and who shall serve for a term of six years unless previously removed by the Secretary of Health, Education, and Welfare.

(c) The Institute is authorized to:

(1) develop and establish recommended occupational safety and health standards; and

(2) perform all functions of the Secretary of Health, Education, and Welfare under sections 20 and 21 of this Act.

(d) Upon his own initiative, or upon the request of the Secretary of Health, Education, and Welfare, the Director is authorized (1) to conduct such research and experimental programs as he determines are necessary for the development of criteria for new and improved occupational safety and health standards, and (2) after consideration of the results of such research and experimental programs make recommendations concerning new or improved occupational safety and health standards. Any occupational safety and health standard recommended pursuant to this section shall immediately be forwarded to the Secretary of Labor, and to the Secretary of Health, Education, and Welfare.

(e) In addition to any authority vested in the Institute by other provisions of this section, the Director, in carrying out the functions of the Institute, is authorized to:

(1) prescribe such regulations as he deems necessary governing the manner in which its functions shall be carried out;

(2) receive money and other property donated, bequeathed, or devised, without condition or restriction other than that it be used for the purposes of the Institute and to use, sell, or otherwise dispose of such property for the purpose of carrying out its functions;

(3) receive (and use, sell, or otherwise dispose of, in accordance with paragraph (2)), money and other property donated, bequeathed, or devised to the Institute with a condition or restriction, including a condition that the Institute use other funds of the Institute for the purposes of the gift;

(4) in accordance with the civil service laws, appoint and fix the compensation of such personnel as may be necessary to carry out the provisions of this section;

(5) obtain the services of experts and consultants in accordance with the provisions of section 3109 of title 5, United States Code;

(6) accept and utilize the services of voluntary and non-compensated personnel and reimburse them for travel expenses, including per diem, as authorized by section 5703 of title 5, United States Code;

(7) enter into contracts, grants or other arrangements, or modifications thereof to carry out the provisions of this section, and such contracts or modifications thereof may be entered into without performance or other bonds, and without regard to section 3709 of the Revised Statutes, as amended (41 U.S.C. 5), or any other provision of law relating to competitive bidding;

(8) make advance, progress, and other payments which the Director deems necessary under this title without regard to the provisions of section 3648 of the Revised Statutes, as amended (31 U.S.C. 529); and

(9) make other necessary expenditures.

(f) The Director shall submit to the Secretary of Health, Education, and Welfare, to the President, and to the Congress an annual report of the operations of the Institute under this Act, which shall include a detailed statement of all private and public funds received and expended by it, and such recommendations as he deems appropriate.

SECTION 23—GRANTS TO THE STATES

(a) The Secretary is authorized, during the fiscal year ending June 30, 1971, and the two succeeding fiscal years, to make grants to the States which have designated a State agency under section 18 to assist them:

(1) in identifying their needs and responsibilities in the area of occupational safety and health,

(2) in developing State plans under section 18, or

(3) in developing plans for:

(A) establishing systems for the collection of information concerning the nature and frequency of occupational injuries and diseases;

(B) increasing the expertise and enforcement capabilities of their personnel engaged in occupational safety and health programs; or

(C) otherwise improving the administration and enforcement of State occupational safety and health laws, including standards thereunder, consistent with the objectives of this Act.

(b) The Secretary is authorized, during the fiscal year ending June 30, 1971, and the two succeeding fiscal years, to make grants to the States for experimental and demonstration projects consistent with the objectives set forth in subsection (a) of this section.
(c) The Governor of the State shall designate the appropriate State agency for receipt of any grant made by the Secretary under this section.
(d) Any State agency designated by the Governor of the State desiring a grant under this section shall submit an application therefore to the Secretary.
(e) The Secretary shall review the application, and shall, after consultation with the Secretary of Health, Education, and Welfare, approve or reject such application.
(f) The Federal share for each State grant under subsection (a) or (b) of this section may not exceed 90 per cent of the total cost of the application. In the event the Federal share for all States under either such subsection is not the same, the differences among the States shall be established on the basis of objective criteria.
(g) The Secretary is authorized to make grants to the States to assist them in administering and enforcing programs for occupational safety and health contained in State plans approved by the Secretary pursuant to section 18 of this Act. The Federal share for each State grant under this subsection may not exceed 50 per cent of the total cost to the State of such a program. The last sentence of subsection (f) shall be applicable in determining the Federal share under this subsection.
(h) Prior to June 30, 1973, the Secretary shall, after consultation with the Secretary of Health, Education, and Welfare, transmit a report to the President and to the Congress, describing the experience under the grant programs authorized by this section and making any recommendations he may deem appropriate.

SECTION 24—STATISTICS

(a) In order to further the purposes of this Act, the Secretary, in consultation with the Secretary of Health, Education, and Welfare, shall develop and maintain an effective program of collection, compilation, and analysis of occupational safety and health statistics. Such program may cover all employments whether or not subject to any other provisions of this Act but shall not cover employments excluded by section 4 of the Act. The Secretary shall compile accurate statistics on work injuries and illnesses which shall include all disabling, serious, or significant injuries and illnesses, whether or not involving loss of time from work, other than minor injuries requiring only first aid treatment and which do not involve medical treatment, loss of consciousness, restriction of work or motion, or transfer to another job.
(b) To carry out his duties under subsection (a) of this section, the Secretary may:

(1) promote, encourage, or directly engage in programs of studies, information and communication concerning occupational safety and health statistics;

(2) make grants to States or political subdivisions thereof in order to assist them in developing and administering programs dealing with occupational safety and health statistics; and

(3) arrange, through grants or contracts, for the conduct of such research and investigations as give promise of furthering the objectives of this section.

(c) The Federal share for each grant under subsection (b) of this section may be up to 50 per cent of the State's total cost.
(d) The Secretary may, with the consent of any State or political subdivision thereof, accept and use the services, facilities, and employees of the agencies of such State or political subdivision, with or without reimbursement, in order to assist him in carrying out his
functions under this section.
(e) On the basis of the records made and kept pursuant to section 8(c) of this Act, employers shall file such reports with the Secretary as he shall prescribe by regulation, as necessary to carry out his functions under this Act.
(f) Agreements between the Department of Labor and States pertaining to the collection of occupational safety and health statistics already in effect on the effective date of this Act shall remain in effect until superseded by grants or contracts made under this Act.

SECTION 25—AUDITS

(a) Each recipient of a grant under this Act shall keep such records as the Secretary or the Secretary of Health, Education, and Welfare shall prescribe, including records which fully disclose the amount and disposition by such recipient of the proceeds of such grant, the total cost of the project or undertaking in connection with which such grant is made or used, and the amount of that portion of the cost of the project or undertaking supplied by other sources, and such other records as will facilitate an effective audit.

(b) The Secretary or the Secretary of Health, Education, and Welfare, and the Comptroller General of the United States, or any of their duly authorized representatives, shall have access for the purpose of audit and examination to any books, documents, papers, and records of the recipients of any grant under this Act that are pertinent to any such grant.

SECTION 26—ANNUAL REPORT

Within one hundred and twenty days following the convening of each regular session of each Congress, the Secretary and the Secretary of Health, Education, and Welfare shall each prepare and submit to the President for transmittal to the Congress a report upon the subject matter of this Act, the progress toward achievement of the purpose of this Act, the needs and requirements in the field of occupational safety and health, and any other relevant information. Such reports shall include information regarding occupational safety and health standards, and criteria for such standards, developed during the preceding year; evaluation of standards and criteria previously developed under this Act, defining areas of emphasis for new criteria and standards; an evaluation of the degree of observance of applicable occupational safety and health standards, and a summary of inspection and enforcement activity undertaken; analysis and evaluation of research activities for which results have been obtained under governmental and non-governmental sponsorship; an analysis of major occupational diseases; evaluation of available control and measurement technology for hazards for which standards or criteria have been developed during the preceding year; description of cooperative efforts undertaken between Government agencies and other interested parties in the implementation of this Act during the preceding year; a progress report on the development of an adequate supply of trained manpower in the field of occupational safety and health, including estimates of future needs and the efforts being made by Government and others to meet those needs; listing of all toxic substances in industrial usage for which labeling requirements, criteria, or standards have not yet been established; and such recommendations for additional legislation as are deemed necessary to protect the safety and health of the worker and improve the administration of this Act.

SECTION 27—NATIONAL COMMISSION ON STATE WORKMEN'S COMPENSATION LAWS

(a)(1) The Congress hereby finds and declares that:

(A) the vast majority of American workers, and their families, are dependent on workmen's compensation for their basic economic security in the event such workers suffer disabling injury or death in the course of their employment; and that the full protection of American workers from job-related injury or death requires an adequate, prompt, and equitable system of workmen's compensation as well as an effective program of occupational health and safety regulation; and

(B) in recent years serious questions have been raised concerning the fairness and adequacy of present workmen's compensation laws in the light of the growth of the economy, the changing nature of the labor force, increases in medical knowledge, changes in the hazards associated with various types of employment, new technology creating new risks to health and safety, and increases in the general level of wages and the cost of living.

(2) The purpose of this section is to authorize an effective study and objective evaluation of State workmen's compensation laws in order to determine if such laws provide an adequate, prompt, and equitable system of compensation for injury or death arising out of or in the course of employment.

(b) There is hereby established a National Commission on State Workmen's Compensation Laws.

(c)(1) The Workmen's Compensation Commission shall be composed of fifteen members to be appointed by the President from among members of State workmen's compensation boards, representatives of insurance carriers, business, labor, members of the medical profession having experience in industrial medicine or in workmen's compensation cases, educators having special expertise in the field of workmen's compensation, and representatives of the general public. The Secretary, the Secretary of Commerce, and the Secretary of Health, Education, and Welfare shall be ex officio members of the Workmen's Compensation Commission:

(2) Any vacancy in the Workmen's Compensation Commission shall not affect its powers.

(3) The President shall designate one of the members to serve as Chairman and one to serve as Vice Chairman of the Workmen's Compensation Commission.

(4) Eight members of the Workmen's Compensation Commission shall constitute a quorum.

(d)(1) The Workmen's Compensation Commission shall undertake a comprehensive study and evaluation of State workmen's compensation laws in order to determine if such laws provide an adequate, prompt, and equitable system of compensation. Such study and evaluation shall include, without being limited to, the following subjects: (A) the amount and duration of permanent and temporary disability benefits and the criteria for determining the maximum limitations thereon, (B) the amount and duration of medical benefits and provisions insuring adequate medical care and free choice of physician, (C) the extent of coverage of workers, including exemptions based on numbers or type of employment, (D) standards for determining which injuries or diseases should be deemed compensable, (E) rehabilitation, (F) coverage under second or subsequent injury funds, (G) time limits on filing claims, (H) waiting periods, (I) compulsory or elective coverage, (J) administration, (K) legal expenses, (L) the feasibility and desirability of a uniform system of reporting information concerning job-related injuries and diseases and the operation of workmen's compensation laws, (M) the resolution of conflict of laws, extraterritoriality and similar problems arising from claims with multistate aspects, (N) the extent to which private insurance carriers are excluded from supplying workmen's compensation coverage and the desirability of such exclusionary practices, to the extent they are found to exist, (O) the relationship between workmen's compensation on the one hand, and old-age, disability, and survivors insurance and other types of insurance, public or private, on the other hand, (P) methods of implementing the recommendations of the Commission.

(2) The Workmen's Compensation Commission shall transmit to the President and to the Congress not later than July 31, 1972, a final report containing a detailed statement of the findings and conclusions of the Commission, together with such recommendations as it deems advisable.

(e)(1) The Workmen's Compensation Commission or, on the authorization of the Workmen's Compensation Commission, any subcommittee or members thereof, may, for the purpose of carrying out the provisions of this title, hold such hearings, take such testimony, and sit and act at such times and places as the Workmen's Compensation Commission deems advisable. Any member authorized by the Workmen's Compensation Commission may administer oaths or affirmations to witnesses appearing before the Workmen's Compensation Commission or any subcommittee or members thereof.

(2) Each department, agency, and instrumentality of the executive branch of the Government, including independent agencies, is authorized and directed to furnish to the Workmen's Compensation Commission, upon request made by the Chairman or Vice Chairman, such information as the Workmen's Compensation Commission deems necessary to carry out its functions under this section.

(f) Subject to such rules and regulations as may be adopted by the Workmen's Compensation Commission, the Chairman shall have the power to:

(1) appoint and fix the compensation of an executive director, and such additional staff personnel as he deems necessary, without regard to the provisions of title 5, United States Code, governing appointments in the competitive service, and without regard to the provisions of chapter 51 and subchapter III of chapter 53 of such title relating to classification and General Schedule pay rates, but at rates not in excess of the maximum rate for GS-18 of the General Schedule under section 5332 of such title, and

(2) procure temporary and intermittent services to the same extent as is authorized by section 3109 of title 5, United States Code.

(g) The Workmen's Compensation Commission is authorized to enter into contracts with Federal or State agencies, private firms, institutions, and individuals for the conduct of research or surveys, the preparation of reports, and other activities necessary to the discharge of its duties.

(h) Members of the Workmen's Compensation Commission shall receive compensation for each day they are engaged in the performance of their duties as members of the Workmen's Compensation Commission at the daily rate prescribed for GS-18 under section 5332 of title 5, United States Code, and shall be entitled to reimbursement for travel, subsistence, and other necessary expenses incurred by them in the performance of their duties as members of the Workmen's Compensation Commission.

(i) There are hereby authorized to be appropriated such sums as may be necessary to carry out the provisions of this section.

(j) On the ninetieth day after the date of submission of its final report to the President, the Workmen's Compensation Commission shall cease to exist.

SECTION 28—ECONOMIC ASSISTANCE TO SMALL BUSINESSES

(a) Section 7(b) of the Small Business Act, as amended, is amended:
(1) by striking out the period at the end of "paragraph (5)" and inserting in lieu thereof "; and"
(2) by adding after paragraph (5) a new paragraph as follows:
"(6) to make such loans (either directly or in cooperation with banks, or other lending institutions through agreements to participate on an immediate or deferred basis) as the Administration may determine to be necessary or appropriate to assist any small business concern in effecting additions to or alterations in the equipment, facilities, or methods of operation of such business in order to comply with the applicable standards promulgated pursuant to section 6 of the Occupational Safety and Health Act of 1970 or standards adopted by a State pursuant to a plan approved under section 18 of the Occupational Safety and Health Act of 1970, if the Administration determines that such concern is likely to suffer substantial economic injury without assistance under this paragraph."
(b) The third sentence of section 7(b) of the Small Business Act, as amended, is amended by striking out "or (5)" after "paragraph (3)" and inserting a comma followed by "(5) or (6)".
(c) Section 4(c)(1) of the Small Business Act, as amended, is amended by inserting "7(b)(6)," after "7(b)(5),".
d) Loans may also be made or guaranteed for the purposes set forth in section 7(b)(6) of the Small Business Act, as amended, pursuant to the provisions of section 202 of the Public Works and Economic Development Act of 1965, as amended.

SECTION 29—ADDITIONAL ASSISTANT SECRETARY OF LABOR

(a) Section 2 of the Act of April 17, 1946 (60 Stat. 91) as amended (29 U.S.C. 553) is amended by:
(1) striking out "four" in the first sentence of such section and inserting in lieu thereof "five"; and
(2) adding at the end thereof the following new sentence, "One of such Assistant Secretaries shall be an Assistant Secretary of Labor for Occupational Safety and Health.".
(b) Paragraph (20) of section 5315 of title 5, United States Code, is amended by striking out "(4)" and inserting in lieu thereof "(5)".

SECTION 30—ADDITIONAL POSITIONS

Section 5108(c) of title 5, United States Code, is amended by:
(1) striking out the word "and" at the end of paragraph (8);
(2) striking out the period at the end of paragraph (9) and inserting in lieu thereof a semicolon and the word "and"; and
(3) by adding immediately after paragraph (9) the following new paragraph:
"(10)(A) the Secretary of Labor, subject to the standards and procedures prescribed by this chapter, may place an additional twenty-five positions in the Department of Labor in GS-16, 17, and 18 for the purposes of carrying out his responsibilities under the Occupational Safety and Health Act of 1970;
"(B) the Occupational Safety and Health Review Commission, subject to the standards and procedures prescribed by this chapter, may place ten positions in GS-16, 17, and 18 in carrying out its functions under the Occupational Safety and Health Act of 1970."

SECTION 31—EMERGENCY LOCATOR BEACONS

Section 601 of the Federal Aviation Act of 1958 is amended by inserting at the end thereof a new subsection as follows:

"EMERGENCY LOCATOR BEACONS

"(d)(1) Except with respect to aircraft described in paragraph (2) of this subsection, minimum standards pursuant to this section shall include a requirement that emergency locator beacons shall be installed:
"(A) on any fixed-wing, powered aircraft for use in air commerce the manufacture of which is completed, or which is imported into the United States, after one year following the date of enactment of this subsection; and
"(B) on any fixed-wing, powered aircraft used in air commerce after three years following such date.

"(2) The provisions of this subsection shall not apply to jet-powered aircraft; aircraft used in air transportation (other than air taxis and charter aircraft); military aircraft; aircraft used solely for training purposes not involving flights more than twenty miles from its base; and aircraft used for the aerial application of chemicals."

SECTION 32—SEPARABILITY

If any provision of this Act, or the application of such provision to any person or circumstance, shall be held invalid, the remainder of this Act, or the application of such provision to persons or circumstances other than those as to which it is held invalid, shall not be affected thereby.

SECTION 33—APPROPRIATIONS

There are authorized to be appropriated to carry out this Act for each fiscal year such sums as the Congress shall deem necessary.

SECTION 34—EFFECTIVE DATE

This Act shall take effect one hundred and twenty days after the date of its enactment.

Approved December 29, 1970.

Amended November 5, 1990.

LEGISLATIVE HISTORY:

HOUSE REPORTS: No. 91-1291 accompanying H.R. 16785 (Comm. on Education and Labor) and No. 91-1765 (Comm. of Conference).
SENATE REPORT: No. 91-1282 (Comm. on Labor and Public Welfare).
CONGRESSIONAL RECORD, Vol. 116 (1970):
Oct. 13, Nov. 16, 17, considered and passed Senate.
Nov. 23, 24, considered and passed House, amended, in lieu of H.R. 16785.
Dec. 16, Senate agreed to conference report.
Dec. 17, House agreed to conference report.

Appendix 2

ACRONYMS AND ABBREVIATIONS

In the practice of occupational safety and health, as in most technical and scientific arenas, numerous acronyms and abbreviations are used quite regularly. The following is a reference listing of some of the most frequently encountered, either in this text or in the safety and health profession in general.

AAAS	American Association for the Advancement of Science
AAEE	American Academy of Environmental Engineers
AAIH	American Academy of Industrial Hygiene
AAOHN	American Association of Occupational Health Nurses
AAP	Asbestos Action Program
ABIH	American Board of Industrial Hygiene
ABOHN	American Board of Occupational Health Nursing
AC	Advisory Circular
ACBM	Asbestos-Containing Building Material
ACEC	American Consulting Engineers Council
ACGIH	American Conference of Governmental Industrial Hygienists
ACM	Asbestos-Containing Material
ACS	American Chemical Society
ACSH	American Council on Science and Health
ADA	American's with Disabilities Act
AEA	Atomic Energy Act
AG	Attorney General
AGA	American Gas Association, Inc.
AHERA	Asbestos Hazard Emergency Response Act
AIChE	American Institute of Chemical Engineers
AIDS	Acquired Immune Deficiency Syndrome
AIHA	American Industrial Hygiene Association
AIHC	American Industrial Health Council
AIHF	American Industrial Hygiene Foundation
AIP	Auto Ignition Point
ALA	American Lung Association
AL	Action Level
ALJ	Administrative Law Judge
AMA	American Medical Association
ANEC	American Nuclear Energy Council

ANPR	Advanced Notice of Proposed Rule Making
ANSI	American National Standards Institute
AO	Administrative Order
APA	Administrative Procedures Act
APHA	American Public Health Association
ASCE	American Society of Civil Engineers
ASCII	American Standard Code for Information Interchange
ASHAA	Asbestos in Schools Hazard Abatement Act
ASHRAE	American Society of Heating, Refrigeration, and Air Conditioning Engineers
ASME	American Society of Mechanical Engineers
ASSE	American Society of Safety Engineers
ASTM	American Society for Testing Materials
ATSDR	Agency for Toxic Substances and Disease Registry
AVD	Alleged Violation Description
BBP	Bureau of Business Practice
BCSP	Board of Certified Safety Professionals
BEI	Biological Exposure Index
BLS	Bureau of Labor Statistics
BNA	Bureau of National Affairs
BOE	Bureau of Explosives
BOM	Bureau of Mines
BTU	British Thermal Unit
C	Celsius, Centigrade (degrees)
CAA	Clean Air Act
CAAA	Clean Air Act Amendments
CAS	Chemical Abstract Service
CBA	Cost-Benefit Analysis
CBU	Collective Bargaining Unit
CC	Cubic Centimeter
CDC	Centers for Disease Control
CEA	Cause and Effect Analysis
CEL	Ceiling Exposure Limit
CERCLA	Comprehensive Environmental Response, Compensation, and Liabilities Act
CEU	Continuing Education Unit
CFC	Chlorofluorocarbons
CFM	Cubic Feet per Minute
CFR	Code of Federal Regulations
CFS	Cubic Feet per Second
CGA	Compressed Gas Association
CHEMTREC	Chemical Transportation Emergency Center

CHRIS	Chemical Hazard Response Information System
CIAQ	Council on Indoor Air Quality
CIH	Certified Industrial Hygienist
CIS	Chemical Information System
CHP	Certified Health Physicist
CMA	Chemical Manufacturers Association
COC	Chain-of-Custody
COHN	Certified Occupational Health Nurse
COHST	Certified Occupational Health and Safety Technologist
COM	Compliance Operations Manual
CPR	Cardio Pulmonary Resuscitation
CPSC	Consumer Product Safety Commission
CSHO	Compliance Safety and Health Officer
CSP	Certified Safety Professional
CSS	Certified Safety Specialist (World Safety Organization)
CTD	Cumulative Trauma Disorder
CTS	Carpal Tunnel Syndrome
dBA	Decibel-A Scale
dB	Decibel
DBT	Dry Bulb Temperature
DHHS	Department of Health and Human Services
DOD	Department of Defense
DOC	Department of Commerce
DOL	Department of Labor
DOJ	Department of Justice
DOT	Department of Transportation
EAJA	Equal Access to Justice Award
EAP	Employee Assistance Program
ECP	Exposure Control Plan
EEC	European Economic Community
EI	Exposure Index
EIS	Environmental Impact Statement
EOP	Emergency Operations Plan
EPA	Environmental Protection Agency
ERT	Emergency Response Team
F	Fahrenheit (degrees)
FACA	Federal Advisory Committee Act
FACM	Friable Asbestos-Containing Materials
FAM	Friable Asbestos Material
FAR	Federal Acquisition Regulations
	Federal Aviation Regulations
FCC	Federal Communications Commission

FDA	Food and Drug Administration (USA)
FEA	Federal Energy Administration
FEMA	Federal Emergency Management Agency
FHA	Fault Hazard Analysis
FIFRA	Federal Insecticide, Fungicide and Rodenticide Act
FLP	Flash Point
FM	Factory Mutual
	Friable Material
FMEA	Failure Mode and Effect Analysis
FOI	Freedom of Information (Act)
FOM	Field Operations Manual
FR	Federal Register
FSF	Flight Safety Foundation
FTA	Fault Tree Analysis
GBT	Globe Temperature
GC/MS	Gas Chromatography/Mass Spectrograph
GFCI	Ground Fault Circuit Interrupter
GIDEP	Government Industry Data Exchange Program
GSA	General Services Administration
GT	Globe Temperature
HA	Hazard Analysis
HAV	Hepatitis A Virus
HAZMAT	Hazardous Materials
HAZWOPER	Hazardous Waste Operations and Emergency Response
HBIG	Hepatitis B Immune Globulin
HBV	Hepatitis B Virus
HC	Hazardous Constituents
HCP	Hazard Communication Program
	Hearing Conservation Program
HE	Hazard Evaluation
HEPA	High Efficiency Particulate Air (filter)
HFS	Human Factors Society
HIV	Human Immunodeficiency Virus
HMIS	Hazardous Materials Information System
HPLC	High Performance Liquid Chromatography
HPS	Health Physics Society
Hz	Hertz
IAP	Indoor Air Pollutant
IARC	International Agency for Research on Cancer
IAQ	Indoor Air Quality
ICRA	Industrial Chemical Research Association

IDLH	Immediately Dangerous to Life and Health
IEEE	Institute of Electrical and Electronics Engineering
IES	Illumination Engineering Society (of North America)
IG	Inspector General
ILO	International Labor Office (United Nations)
ISEA	International Safety Equipment Association
ISO	International Organization for Standards
JAG	Judge Advocate General (Military)
JSA	Job Safety Analysis
JHA	Job Hazard Analysis
keV	Kilovolts (radiation)
kHz	Kilohertz
kWH	Kilowatt Hour
KVP	Kilovoltage Peak
LASER	Light Amplification by Stimulated Emission of Radiation
LBP	Lead-Based Paint
LC	Lethal Concentration
LCD	Liquid Crystal Display
LD	Lethal Dose
LEL	Lower Explosive Limit
LFL	Lower Flammability Limit
LOD	Level of Detection
LPG	Liquid Propane Gas
LPN	Licensed Practical Nurse
LRMS	Low Resolution Mass Spectroscopy
LSC	Life Safety Code®
LWDI	Lost Work-Day Injury
LVN	Licensed Vocational Nurse
MeV	Million Electron Volts
MED	Minimum Effective Dose
mg/m^3	Milligrams per Cubic Meter
MIL-STD	Military Standard
MMWR	Morbidity and Mortality Weekly Report
MORT	Management Oversight and Risk Tree
MPC	Maximum Permissible Concentration
MPD	Maximum Permissible Dose
MPE	Maximum Permissible Exposure
MP	Melting Point
MREM	Millirem
MSDS	Material Safety Data Sheet

MSHA	Mine Safety and Health Administration
MT	Metric Ton
mW	Milliwatt
MW	Microwave
NAR	National Asbestos Registry
NASHP	National Association of Safety and Health Professionals
NASA	National Aeronautics and Space Administration
NBS	National Bureau of Standards
NCAQ	National Commission on Air Quality
NCI	National Cancer Institute
NCRP	National Council on Radiation Protection and Measurement
NEC	National Electrical Code®
NEHA	National Environmental Health Association
NEMA	National Electrical Manufacturers Association
NFPA	National Fire Protection Association
NIDA	National Institute for Drug Abuse
NIH	National Institutes of Health
NIOSH	National Institute for Occupational Safety and Health
NIST	National Institute of Standards and Technology
NOHS	National Occupational Health Survey
NPL	National Priorities List
NPR	Notice of Proposed Rulemaking
NRC	Nuclear Regulatory Commission
NRR	Noise Reduction Rating
NSC	National Safety Council
NSMS	National Safety Management Society
NTP	National Toxicology Program
NTSB	National Transportation Safety Board
ODC	Ozone Depleting Chemical
OHA	Operating Hazard Analysis
OHN	Occupational Health Nurse
OLDS	Ozone Level Depleting Substances
O&SHA	Operating and Support Hazard Analysis
OSHA	Occupational Safety and Health Administration
OSHAct	Occupational Safety and Health Act (of 1970)
OSHRC	Occupational Safety and Health Review Commission
OTS	Other-Than-Serious (violation category)
PAPR	Powered Air Purifying Respirator
PbB	Blood Lead
PC	Potential Carcinogen

PCB	Polychlorinated Biphenyl
PDC	Professional Development Conference
PEL	Permissible Exposure Limit (OSHA)
PHA	Preliminary Hazard Analysis
	Process Hazard Analysis
PHSA	Public Health Service Act
PL	Public Law
PMA	Petition for the Modification of Abatement
POC	Point of Contact
PPB	Parts per Billion
PPE	Personal Protective Equipment
PPM	Parts Per Million
PSIA	Pounds Per Square Inch (Absolute)
PSI	Pounds Per Square Inch
PSIG	Pounds Per Square Inch (Gage)
PSTD	Post Traumatic Stress Disorder
PTS	Permanent Threshold Shift
QA	Quality Assurance
QC	Quality Control
RAC	Radiation Advisory Committee
	Risk Assessment Code
RBE	Relative Biological Effectiveness
RCRA	Resource Conservation and Recovery Act
REC	Recognize, Evaluate, and Control
REL	Recommended Exposure Limit (NIOSH)
REM	Roentgen Equivalent Man
RF	Radiofrequency
RFP	Request for Proposal
RH	Relative Humidity
RN	Registered Nurse
RTECS	Registry of Toxic Effects of Chemical Substances
RQ	Reportable Quantity
SARA	Superfund Amendments and Reauthorization Act
SAVE	Standard Alleged Violation Element
SCBA	Self-Contained Breathing Apparatus
SCFM	Standard Cubic Feet per Minute
SG	Surgeon General
SHA	System Hazard Analysis
SIC	Standard Industry Classification
SLM	Sound Level Meter
SOP	Standard Operating Procedure

SOW	Statement of Work
SPF	Single Point Failure
SSA	System Safety Analysis
SSS	System Safety Society
STEL	Short-term Exposure Limit
STS	Standard Threshold Shift
TB	Tuberculosis
TD	Toxic Dose
TLV	Threshold Limit Value (ACGIH)
TLV-C	Threshold Limit Value - Ceiling
TLV-skin	Threshold Limit Value - Skin Absorption
TLV-STEL	Threshold Limit Value - Short-term Exposure Limit
TQM	Total Quality Management
TQMS	Total Quality Management System
TSCA	Toxic Substances Control Act
TTS	Temporary Threshold Shift
TWA	Time-weighted Average
UEL	Upper Explosive Limit
UFL	Upper Flammability Limit
UL	Underwriter's Laboratory
UN	United Nations
USC	United States Code
UV	Ultra-violet
V	Volt
VC	Vital Capacity
VDT	Video Display Terminal
VOC	Volatile Organic Compound
VOS	Veterans of Safety
VP	Vapor Pressure
VPP	Voluntary Protection Program
WBGT	Wet Bulb Globe Temperature (or "Test")
WBT	Wet Bulb Temperature
WHO	World Health Organization
WSO	World Safety Organization

Appendix 3

OSHA Offices

The following is a listing of the OSHA National, Regional, and Area Offices, current as of December 1996.

National Office:
U.S. Department of Labor
Occupational Safety and Health Administration
3rd & Constitution Avenue, N.W.
Washington, DC 20210

Regional Offices:

Region I (CT, ME, MA, NH, RI, VT)
16-18 North Street
1 Dock Square Building, 4th Floor
Boston, MA 02109
(617) 565-1145

Region II (NY, NJ, PR)
201 Varick Street
New York, NY 10014
212-337-2378

Region III (DE, DC, MD, PA, VA, WV)
Gateway Building, Suite 2100
3535 Market Street
Philadelphia, PA 19104
(215) 596-1201

Region IV (AL, FL, GA, KY, MS, NC, SC, TN)
1375 Peachtree Street, N.E., Suite 587
Atlanta, GA 30367
(404) 347-3573

Region V (IL, IN, MN, MI, OH, WI)
230 South Dearborn Street
32nd Floor, Room 3244
Chicago, IL 60604
(313) 353-2220

Region VI (AR, LA, NM, OK, TX)
525 Griffin Square, Room 602
Dallas, TX 75202
(214) 767-4731

Region VII (IA, KS, MO, NE)
911 Walnut Street, Room 406
Kansas City, MO 64106
(816) 374-5861

Region VIII (CO, MT, ND, SD, UT, WY)
Federal Building, Room 1576
1961 Stout Street
Denver, CO 80294
(303) 844-3061

Region IX (AZ, CA, HI, NV)
71 Stevenson Street, 4th Floor
San Francisco, CA 94105
(415) 995-5672

Region X (AK, ID, OR, WA)
Federal Office Building, Room 6003
909 First Avenue
Seattle, WA 98174
(206) 442-5930

Area Offices:

Alabama
Occupational Safety and Health
Administration
2047 Canyon Road - Todd Mall
Birmingham, AL 35216
(205) 731-1534

Alaska
Occupational Safety and Health
Administration
Federal Building
701 "C" Street, Box 79
Anchorage, AK 99513
(907) 271-5152

Arizona
Occupational Safety and Health
Administration
3221 North 16th Street, Suite 100
Phoenix, AZ 85016
(602) 241-2006

Arkansas
Occupational Safety and Health
Administration
Savers Building, Suite 828
320 West Capitol Avenue
Little Rock, AR 72201
(501) 378-6291

California
Occupational Safety and Health
Administration
400 Oceangate, Suite 530
Long Beach, CA 90802
(213) 514-6387

Occupational Safety and Health
Administration
2422 Arden Way, Suite A-1
Sacramento, CA 95825
(916) 646-9220

Occupational Safety and Health
Administration
7807 Convoy Court, Suit 160
San Diego, CA 92111
(619) 569-9071

California *(continued)*
Occupational Safety and Health
Administration
950 South Bascom, Suite 3120
San Jose, CA 95128
(408) 291-4600

Occupational Safety and Health
Administration
801 Ygnacio Valley Road, Room 205
Walnut Creek, CA 94596-3823
(415) 943-1973

Occupational Safety and Health
Administration
100 North Citrus Avenue, Suite 240
West Covina, CA 91791
(818) 915-1558

Colorado
Occupational Safety and Health
Administration
Tremont Center, 1st Floor
333 West Colfax
Denver, CO 80204
(303) 844-5285

Connecticut
Occupational Safety and Health
Administration
Federal Office Building
450 Main Street
Room 508
Hartford, CT 06103
(203) 240-3152

Florida
Occupational Safety and Health
Administration
299 East Broward Boulevard
Fort Lauderdale, FL 33301
(305) 527-7292

Occupational Safety and Health
Administration
3100 University Boulevard South
Jacksonville, FL 32207
(904) 791-2895

Florida *(continued)*
Occupational Safety and Health
Administration
5807 Brekenridge Parkway
Suite A
Tampa, FL 33610
(813) 626-1177

Georgia
Occupational Safety and Health
Administration
Building 10, Suite 33
La Vista Perimeter Office Park
Tucker, GA 30084
(404) 331-4767

Hawaii
Occupational Safety and Health
Administration
300 Ala Moana Boulevard, Suite 5122
P.O. Box 50072
Honolulu, HI 96850
(808) 541-2685

Idaho
Occupational Safety and Health
Administration
Federal Building/USCH, Room 324
550 West Fort Street, Box 007
Boise, ID 83724
(208) 334-1867

Illinois
Occupational Safety and Health
Administration
344 Smoke Tree Business Park
North Aurora, IL 60542
(312) 896-8700

Occupational Safety and Health
Administration
1600 167th Street, Suite 12
Calumet City, IL 60409
(312) 891-3800

Occupational Safety and Health
Administration
6000 West Touhy Avenue
Niles, IL 60648
(312) 631-8200

Illinois *(continued)*
Occupational Safety and Health
Administration
2001 West Willow Knolls Road, Suite 101
Peoria, IL 61614
(309) 671-7033

Indiana
Occupational Safety and Health
Administration
U.S. Post Office and Courthouse
46 East Ohio Street, Room 423
Indianapolis, IN 46204
(317) 269-7290

Iowa
Occupational Safety and Health
Administration
210 Walnut Street, Room 815
Des Moines, IA 50309
(515) 284-4794

Kansas
Occupational Safety and Health
Administration
216 North Waco, Suite B
Wichita, KS 67202
(316) 269-6644

Kentucky
Occupational Safety and Health
Administration
John C. Watts Federal Building, Room 108
330 West Broadway
Frankfort, KY 40601
(502) 227-7024

Louisiana
Occupational Safety and Health
Administration
2156 Wooddale Boulevard
Hoover Annex, Suite 200
Baton Rouge, LA 70806
(504) 389-0474

Maine
Occupational Safety and Health Administration
U.S. Federal Building
40 Western Avenue, Room 121
Augusta, ME 04330
(207) 622-8417

Maryland
Occupational Safety and Health Administration
Federal Building, Room 1110
Charles Center, 31 Hopkins Plaza
Baltimore, MD 21201
(301) 962-2840

Massachusetts
Occupational Safety and Health Administration
1550 Main Street, Room 532
Springfield, MA 01103-1493
(413) 785-0123

Occupational Safety and Health Administration
400-2 Totten Pond Road, 2nd Floor
Waltman, MA 02154
(617) 647-8681

Michigan
Occupational Safety and Health Administration
300 East Michigan Avenue, Room 202
Lansing, MI 48933
(517) 377-1892

Minnesota
Occupational Safety and Health Administration
110 South 4th Street, Room 425
Minneapolis, MN 55401
(612) 348-1994

Mississippi
Occupational Safety and Health Administration
Federal Building, Suite 1445
100 West Capitol Street
Jackson, MS 39269
(601) 965-4606

Missouri
Occupational Safety and Health Administration
911 Walnut Street, Room 2202
Kansas City, MO 64106
(816) 374-2756

Occupational Safety and Health Administration
4300 Goodfellow Boulevard
Building 105E
St. Louis, MO 63120
(314) 263-2749

Montana
Occupational Safety and Health Administration
19 North 25th Street
Billings, MT 59101
(406) 657-6649

Nebraska
Occupational Safety and Health Administration
Overland-Wolf Building, Room 100
6910 Pacific Street
Omaha, NE 68106
(402) 221-3182

New Hampshire
Occupational Safety and Health Administration
Federal Building, Room 334
55 Pleasant Street
Concord, NH 03301
(603) 225-1639

New Jersey
Occupational Safety and Health Administration
Plaza 35, Suite 205
1030 Saint Georges Avenue
Avenel, NJ 07001
(201) 750-3270

Occupational Safety and Health Administration
2101 Ferry Avenue, Room 403
Camden, NJ 08104
(609) 757-5181

New Jersey *(continued)*
Occupational Safety and Health
Administration
2 East Blackwell Street
Dover, NJ 07801
(201) 361-4050

Occupational Safety and Health
Administration
500 Route 17 South, 2nd Floor
Hasbrouck Heights, NJ 07604
(201) 288-1700

New Mexico
Occupational Safety and Health
Administration
320 Central Avenue, S.W., Suite 13
Albuquerque, NM 87102
(505) 766-3411

New York
Occupational Safety and Health
Administration
Leo W. O'Brien Federal Building
Clinton Avenue and North Pearl Street
Room 132
Albany, NY 12207
(518) 472-6085

Occupational Safety and Health
Administration
5360 Genesee Street
Bowmansville, NY 14026
(716) 684-3891

Occupational Safety and Health
Administration
136-21 Roosevelt Avenue
Flushing, NY 11354
(718) 445-5005

Occupational Safety and Health
Administration
90 Church Street, Room 1405
New York, NY 10007
(212) 264-9840

New York *(continued)*
Occupational Safety and Health
Administration
100 South Clinton Street, Room 1267
Syracuse, NY 13260
(315) 423-5188

Occupational Safety and Health
Administration
990 Westbury Road
Westbury, NY 11590
(516) 334-3344

North Carolina
Century Station, Room 104
300 Fayetville Street Mall
Raleigh, NC 27601
(919) 856-4770

North Dakota
Occupational Safety and Health
Administration
Federal Building, Room 348
P.O. Box 2439
Bismark, ND 58501
(701) 255-4011, ext. 521

Ohio
Occupational Safety and Health
Administration
Federal Office Building, Room 4028
550 Main Street
Cincinnati, OH 45202
(513) 684-3784

Occupational Safety and Health
Administration
Federal Office Building, Room 899
1240 East Ninth Street
Cleveland, OH 44199
(216) 522-3818

Occupational Safety and Health
Administration
Federal Office Building, Room 634
200 North High Street
Columbus, OH 43215
(614) 469-5582

Ohio *(continued)*
Occupational Safety and Health Administration
Federal Office Building, Room 734
234 North Summit Street
Toledo, OH 43604
(419) 259-7542

Oklahoma
Occupational Safety and Health Administration
420 West Main Place, Suite 725
Oklahoma City, OK 73102
(405) 231-5351

Oregon
Occupational Safety and Health Administration
1220 S.W. Third Street, Room 640
Portland, OR 97204
(503) 221-2251

Pennsylvania
Progress Plaza
49 North Progress Avenue
Harrisburg, PA 17109
(717) 782-3902

Occupational Safety and Health Administration
U.S. Custom House, Room 242
Second and Chestnut Street
Philadelphia, PA 19106

Occupational Safety and Health Administration
1000 Liberty Avenue, Room 2236
Pittsburgh, PA 15222
(412) 644-2903

Occupational Safety and Health Administration
Penn Place, Room 2005
20 North Pennsylvania Avenue
Wilkes-Barre, PA 18701
(717) 826-6538

Puerto Rico
Occupational Safety and Health Administration
U.S. Courthouse & Federal Office Building
Carlos Chardon Avenue, Room 559
Hato Rey, PR 00918
(809) 753-4457/4072

Rhode Island
Occupational Safety and Health Administration
380 Westminster Mall, Room 243
Providence, RI 02903
(401) 528-4669

South Carolina
Occupational Safety and Health Administration
1835 Assembly Street, Room 1468
Columbia, SC 29201
(803) 765-5904

Tennessee
Occupational Safety and Health Administration
2002 Richard Jones Road, Suite C-205
Nashville, TN 37215
(615) 736-5313

Texas
Occupational Safety and Health Administration
611 East 6th Street, Room 303
Austin, TX 78701
(512) 482-5783

Occupational Safety and Health Administration
Government Plaza, Room 300
400 Main Street
Corpus Christi, TX 78401
(512) 888-3257

Occupational Safety and Health Administration
2320 La Branch Street, Room 1103
Houston, TX 77004
(713) 750-1727

Texas *(continued)*
Occupational Safety and Health
Administration
1425 West Pioneer Drive
Irving, TX 75061
(214) 767-5347

Occupational Safety and Health
Administration
Federal Building, Room 422
1205 Texas Avenue
Lubbock, TX 79401
(806) 743-7681

Utah
Occupational Safety and Health
Administration
1781 South 300 West
Salt Lake City, UT 84115
(801) 524-5080

Washington
Occupational Safety and Health
Administration
121 107th Street, N.E.
Bellevue, WA 98004
(206) 442-7520

West Virginia
550 Eagan Street, Room 206
Charleston, WV 25301
(304) 347-5937

Wisconsin
Occupational Safety and Health
Administration
2618 North Ballard Road
Appleton, WI 54915
(414) 734-4521

Occupational Safety and Health
Administration
2934 Fish Hatchery Road, Suite 225
Madison, WI 53713
(608) 264-5388

Occupational Safety and Health
Administration
Henry S. Reuss Building, Suite 1180
310 West Wisconsin Avenue
Milwaukee, WI 53203
(414) 291-3315

Glossary of Terms

The following terms are commonly encountered in the occupational safety and health profession. The definitions provided here are intended to facilitate understanding of these terms in relation to the compliance process.

abatement. The removal or elimination of a hazard or the actions taken to effect same.

accident. An unplanned and therefore unwanted or undesired event which results in physical harm and/or property damage.

accident analysis. A concerted, organized, methodical, planned process of examination and evaluation of all evidence identified during investigation.

accident investigation. A methodical effort to collect and interpret facts; a systematic look at the nature and extent of the accident, the risks taken, and loss(es) involved; an inquiry as to how and why the accident event occurred.

accident potential. A situation comprised of human behaviors and/or physical conditions having a probability of resulting in an accident.

accident risk. A measure of vulnerability to loss, damage, or injury caused by a dangerous element or factor (MIL-STD-1574A).

achievable duty. Term used to describe OSHA's approach to employer compliance with safety and health regulations and standards. Contention is that compliance must be achievable within the feasible bounds of economics and technology.

acoustics. The science of sound, including its production, transmission, and effects.

Acquired Immune Deficiency Syndrome (AIDS). A severe (life threatening) disease which represents the late clinical stage of infection with the *human* *immunodeficiency virus (HIV).* The HIV most often results in progressive damage to the immune system and various organ systems, especially the central nervous system. Body fluid-to-body fluid contact with an infected HIV carrier is required for transmission. HIV has been recovered from body fluids other than blood, such as tears, saliva, urine, bronchial secretions, spinal fluid, feces, vomitus, and others.

action level. A exposure limit usually set at 50 percent of the permissible exposure limit (PEL) as specified by the applicable Occupational Safety and Health Administration (OSHA) Standard. Exposures exceeding the action level typically require implementation of certain actions, such as medical surveillance, training, and monitoring programs, but not necessarily further controls (e.g., engineering controls) aimed at reducing exposures.

acute exposure. *Chemical.* A sudden, short, rapid association with a chemical compound. *Radiation.* Exposure of short duration, generally taken to be the total dose absorbed within 24 hours. *Biologic.* A brief encounter with a pathogenic or nonpathogenic microorganism.

acute toxicity. The ability of a substance to cause poisonous effects resulting in severe biological harm or death soon after a single exposure or dose. Also, any severe poisonous effect resulting from a single short-term exposure to a toxic substance.

adjudication. To hear and decide a case; refers to the judge's decision.

administrative control. A measure initiated to reduce worker exposure to various stresses in the work environment. An example is limiting the amount of time an employee can work around health hazards.

administrative law. Refers to that body of law that governs the methods by which administrative agencies make and implement decisions. Federal administrative law is based primarily in specific provisions of the U.S. Constitution, as well as in various other federally mandated statutes. It is within this regulatory framework of administrative law that the basis for occupational safety and health legislation obtains the force of law.

admission. A granting of truth or a conceding of truth from the courts.

advisory committees. Groups of experts used by OSHA to study and advise on certain regulatory issues. The committees consist of persons from outside the agency who may have certain expertise in a given area. These committees do not supersede OSHA's regulatory powers or responsibilities. They only provide input on specific technical and/or policy issues that arise in the course of agency activities.

affirmative defense. A category of defending against an OSHA citation which basically holds that the employer is not specifically arguing the fact that the cited condition(s) existed. The defense is really to the contrary. By not disputing the cite itself, the employer actually affirms the allegation of non-compliance but offers substantial proof to justify reasons for not complying with the cited standard(s).

AIDS. See *Acquired Immune Deficiency Syndrome.*

air-purifying respirator. A device worn by an individual that filters the air to be breathed and returns it to an acceptable state. Also known as *chemical cartridge respirators,* this type of respirator relies on the person's own breathing force to draw air through the filter medium, or it may utilize a powered blower to provide breathing air (i.e., a powered air-purifying respirator or PAPR).

amelioration. Improvement of conditions immediately after an accident; the immediate treatment of injuries and conditions which endanger people and/or property.

anthropometric Relating to human body measurements and modes of action to determine their influence on the safe and efficient operation of equipment.

asphyxiation The depravation of oxygen caused by chemical or physical means. Chemical asphyxiants prevent oxygen transfer from the blood to the body cells. Physical asphyxiants prevent oxygen from reaching the blood.

asbestos. A generic term used to describe a number of naturally occurring fibrous, hydrated mineral silicates differing in chemical composition. They are white, gray, green, or brown. Asbestos fibers are characterized by high tensile strength, flexibility, heat and chemical resistance, and good frictional properties. Chrysotile, crocidolite, amosite, anthophyllite, and actinolite are all forms of asbestos. Exposure to asbestos fibers is known to cause a variety of diseases, including *asbestosis* (a diffuse, interstitial non-malignant scarring of the lung tissues), *bronchogenic carcinoma* (a lung cancer), *mesothelioma* (a tumor of the lining of the chest cavity or lining of the abdomen), and *cancer* of the stomach, colon, and rectum.

Asbestos-Containing Materials (ACM). Any materials containing asbestos which can be released upon destruction or disturbance of the structural integrity of the material.

asbestosis. Disease caused by prolonged exposure to asbestos fibers, which injure the lungs and diminish their oxygen absorbing properties, thus reducing lung function (see *asbestos*).

audible sound. Sound containing frequency components between 20 and 20,000 Hz.

audiogram. A graphic recording of hearing levels referenced to a normal sound pressure level as a function of frequency. Audiograms are used in the diagnosis and treatment of hearing loss (see *audiometer*).

audiometer. A frequency-controlled audio signal generator that produces pure tones at various frequencies and intensities and that is used to measure hearing sensitivity or acuity. Measurement of hearing threshold results in an audiogram, measured in decibels at selected frequencies (see *audiogram*).

audiologist. A professional, specializing in the study and rehabilitation of hearing, who is certified by the American Speech-Language-Hearing Association or licensed by a state board of examiners (OSHA).

aural insert protectors. A form of hearing protector commonly known as earplugs. They are available in numerous configurations as foam, plastic, fine glass fiber, and wax-impregnated cotton. The three types are formable, custom-molded, and premolded.

background noise. Noise which is coming from sources other than that which is being measured.

baseline audiogram. The audiogram against which future audiograms are compared (OSHA).

biohazard. Term applied to organisms or products of organisms that present a health risk to humans. Derived from a combination of the words *biological* and *hazard.*

biohazard area. Any area in which work has been or is being performed with biohazardous agents or materials.

Biological Exposure Index (BEI). Set of reference values established by the American Conference of Government Industrial Hygienists as guidelines for the evaluation of potential health hazards in biological specimens collected from a healthy worker who has been exposed to chemicals to the same extent as a worker with inhalation exposure to the threshold limit value. The values apply to 8-hour exposures, 5 days per week.

biological half-life. The time required for the body to eliminate one-half of an administered dose of any substance by regular process of elimination. This time is approximately the same for both stable and radioactive isotopes of a particular element.

biological hazard. See *biohazard.*

bloodborne pathogens. Pathogenic microorganisms that are present in human blood and can cause disease in humans. These pathogens include, but are not limited to, hepatitis B virus (HBV) and the human immune deficiency virus (HIV).

byssinosis. A disease of the lungs caused by chronic exposure to cotton dust.

catastrophic. A loss of extraordinary magnitude in physical harm to people or damage and destruction of property.

carcinogen. A substance known to cause cancer in humans and animals representing a broad range of organic and inorganic chemicals, hormones, immuno-suppresants, and solid-state materials.

carcinogenic. Describes agents known to induce cancers.

carcinoma. Malignant neoplasm composed of epithelial cells, regardless of the derivation.

Carpal Tunnel Syndrome (CTS). A cumulative trauma disorder (CTD) often associated with activities involving flexing or extending the wrists, or repeated force on the base of the palm and wrist. The carpal tunnel is an opening in the wrist under the carpal ligament on the palmar side of the carpal bones in the wrist. The median nerve, the finger flexor tendons, and blood vessels all pass through this tunnel. Overuse of the tendons can cause them to become inflamed and swollen, creating pressure against the adjacent median nerve and resulting in CST. Symptoms include tingling, pain, or numbness in the thumb and first three fingers.

Ceiling Exposure Limit (CEL). The concentration of a chemical to which workers should not be exposed, even instantaneously, during any part of a working day. Also referred to as *ceiling level,* or *CE.*

chain-of-custody. In legal terms, regulatory agencies as well as employers must be able to verify the chain of possession and custody of any physical samples (air, water, soil, biological, etc.,) that may be used to support litigation. Procedures to ensure this chain-of-custody include written records that can be used to trace possession and handling of the sample from its point of origin through analysis and its introduction as evidence. Without a continuous record of chain-of-custody, the validity of any sample or the results of any tests/analyses may be questioned.

chemical cartridge respirator. An air-purifying respirator capable of filtering out chemical contaminants from the air that is breathed. It usually acts by the use of chemical sorbant pads encased in cartridges that are attached to the respirator facepiece.

chemical compound. A substance composed of two or more elements combined in a fixed and definite proportion by weight.

chronic exposure. (1) *Chemical.* Continual exposure to low levels of a chemical over a long period of time (usually 3 years or more), which can produce symptoms and disease. (2) *Radiation.* Exposure to radiation for long duration by fractionation or protraction. Generally any dosage absorbed over a 24-hour period or longer.

chronic toxicity. The capacity of a substance to cause long-term poisonous human health effects.

commerce clause. A provision in the U.S. Constitution granting Congress the power to authorize administrative agencies such as OSHA to act. Specifically, the commerce clause grants Congress the power *"to regulate commerce among the States."*

Code of Hammurabi. Ancient code of the Babylonian Empire (circa 2100 B.C.) used to prescribe punishment to those who were found to be at fault in the cause of an accident which resulted in the injury or death of another worker. It was the literal implementation of the *eye-for-an-eye* approach to social justice.

common authority. Refers to that entity at a multi-employer worksite usually the general contractor or owner) who has the authority to permit entry of an OSHA representative to conduct an inspection thereby waiving any right or expectation of privacy other contractors working on the property may have.

compliance plan. A documented approach to OSHA compliance required by some specific standards, such as Hazard Communication, which essentially establishes the employer's intended methods of achieving compliance.

consensus standard. A standard of conduct that has been developed by a nationally recognized organization having understood expertise in a given field (such as the American National Standards Institute and the National Fire Protection Association). Such standards carry no force of law unless adopted and implemented by a regulatory agency such as OSHA.

confined space. Any space not designed or intended for continuous occupancy that has a limited or restricted means of entry or exit, and that is subject to the accumulation of toxic or flammable contaminants or has an oxygen-deficient atmosphere. Confined spaces must be large enough for an employee to enter and perform assigned work. Where confined spaces are categorized, there are two levels of classification: permit-required confined space and low-hazard permit space.

cumulative trauma disorder. A collective term used to describe syndromes characterized by discomfort, impairment, disability, or persistent pain in the joints, muscles, tendons, and other soft tissues, with or without physical manifestations. It is often caused, precipitated, or aggravated by repetitive or forced motions which may occur in many differed occupational activities.

danger. Term of warning applied to a condition, operation, or situation that has the potential for physical harm to personnel and/or damage to property.

dBA. Refers to decibels measure on the A scale which is a frequency weighting network that approximates the response of the human ear.

decibel (dB). A non-dimensional unit of measure used to express sound levels. It is a logarithmic expression of the ratio between a measured sound and a reference level of sound. The commonly used "A" scale of measurement, expressed as dBA, is a sound level that approximates the sensitivity of the human ear.

defect. Substandard physical condition, either inherent in the material or created through another action or event.

delegation doctrine. A principle of Constitutional law based upon the classic understanding that Congress, as the duly elected representative of the people, is the repository of all legislative power. Only the people can grant this power to the Congress. According to the delegation doctrine, Congress can not, in turn, delegate this legislative power to another party, such as an administrative agency, because the agency has not been elected by the people. Under strict application of this doctrine, Congress is required to provide reasonably clear and specific statutory standards to guide agency decision making.

de minimis violation. As defined in the OSHAct, a violation which has no direct or immediate relationship to safety or health.

designated representative. Any person or part acting on behalf of an employee with the full permission of that employee. Can include labor unions, relatives, and attorneys. Under certain circumstances, designated representatives may have access to employee exposure and medical records.

disability. An impairment or defect of one or more organs or body members. For *worker's compensation* purposes, the following categories are generally used to determine the level of benefit to be awarded: (1) *Permanent partial.* A permanent physical impairment (loss of an eye, hand, etc.) that restricts the ability of the worker to perform certain jobs. Benefits are normally based upon the percentage of disability incurred. (2) *Permanent total.* A disability that is so extensive it prevent the worker from obtaining or competing for a job. (3) *Temporary partial.* A condition that leaves the employee capable of performing some work and will probably improve to pre-injury or illness status over time and with treatment. (4) *Temporary total.* A disability that renders the worker incapable of working, but from which he or she is expected to recover fully.

disease. A deviation from normal health status associated with a characteristic sequence of signs and symptoms and caused by a specific etiologic agent.

dual capacity doctrine. If an employer acts in a capacity other than that strictly of employer in a "dual capacity State" and an employee is injured, then the employer may be sued for negligence arising out of its dual capacity role.

earplugs. See *aural insert protectors.*

ear protector. Any device worn by an individual to guard against hearing loss in high-noise environments. Some devices may also be used to protect against cold exposures and to prevent the entry of water into the ears.

Egregious Policy. OSHA's fining strategy implemented in 1990 which allowed the agency to fine employers for multiple violations of the same standard as if each were a separate and distinct violation. This allowed the assessment of huge fines against employers found to be in violation of the same requirement in several different instances during an OSHA inspection.

Emergency Temporary Standard (ETS). See *Section 6 (c) standard.*

employee. The person taking the direction from the employer.

employer. The person who has the authority to direct and control the activities of another. Also, the person who supervises the employee on a day to day basis is usually considered the employer. This means that temporary and part-time workers may be considered "employees."

engineering controls. Measures taken to prevent or minimize hazard exposure through the application of controls such as improved ventilation, noise reduction techniques, chemical substitution, equipment and facility modifications, etc.

environmental health. The body of knowledge concerned with the prevention of disease through the control of biological, chemical, or physical agents in air, water, and food. Also concerned with the control of environmental factors that may have an impact on the well-being of people.

ergonomics. A multi-disciplinary activity that concentrates on the interactions between the human and their total working environment with consideration for the stressers that may be present in that environment such as atmospheric heat, illumination, and sound as well as all the tools and equipment used in the work place. Also referred to as *human factors* and *human factors engineering.*

establishment. For the purpose of maintaining records, an establishment is considered a single physical location where business is conducted or where services or industrial operations are performed (for example: a factory, mill, store, hotel, restaurant, movie theater, farm, ranch, bank, sales office, warehouse, or central administrative office).

establishment list. A list that contains the names of particular plants located within the territorial jurisdiction of the local OSHA Area Office that are of the types of industries that have been noted on the industry rank report. See also *industry rank report.*

executive branch of government. That branch of government consisting of the chief executive (i.e., the President), and those offices and positions held under its control.

experience rating. A method of adjusting worker's compensation rates using a three-year history of the employer's claim experience.

expert testimony The opinion of a person skilled in a particular art, science, or profession, having demonstrated special knowledge through experience and education, beyond that which is normally considered common for that art, science, or profession.

expert witness. A witness qualified as a subject expert based upon their knowledge, skill, experience, training, or education. Unlike other witnesses, an expert's testimony may be in the form of an opinion.

explosion. A rapid build-up and release of pressure caused by chemical reaction or by an over-pressurization with a confined space leading to a massive rupture of the pressurized container.

exposure. (1) The amount of radiation or pollutant that represents a potential health risk. (2) The opportunity of a susceptible host to acquire an infection by either direct or indirect mode of transmission. (3) The amount of biological, physical, or chemical agent that reaches a target population. (4) The route by which an organism comes in contact with a toxicant (inhalation, ingestion, dermal absorption, injection).

eye protector A device worn by a person or affixed to equipment to deter harmful substances from contact with the human eye.

feasibility study. Performed by OSHA to determine if a proposed standard is practical for the exposure under consideration as well as from an implementation perspective.

Federal Register. The official daily publication of the United States government that provides a uniform system for publishing presidential and federal agency documents.

first-aid. Minor scratches, burns, splinters, and so forth, which do not ordinarily require medical care, are considered to be first-aid injuries only. Such one-time treatment and follow-up, even if conducted by a physician, are still considered first-aid in nature.

flash point. The minimum temperature at which a flammable liquid gives-off sufficient vapors in the air immediately above the surface of that liquid to ignite if an ignition source is introduced into that area. The lower the flash point, the more flammable the liquid.

formal rulemaking. The process of promulgating rules based upon the formal procedures established in the Administrative Procedures Act (APA) of 1946 requiring (most notably) hearings, substantiation of evidence, and the cross-examination of witnesses.

friable. Refers to materials that have a tendency to crumble easily. Most often used to describe the condition that exists when asbestos fibers can potentially be released and become airborne presenting a respiratory hazard.

general duty clause Refers to Section 5(a)(1) of the Occupational Safety and Health Act of 1970, which states: "Each employer shall furnish to each of his employees employment and a place of employment which are free from recognized hazards that are causing or are likely to cause death or serious physical harm to his employees, and shall comply with the occupational safety and health standards promulgated under this Act."

goggle. A tight-fitting device worn over the eyes to provide splash and/or impact protection.

greater hazard defense. A well-established OSHRC doctrine that, on some occasions, allows employers to escape sanctions for violations of otherwise applicable safety regulations because the act of abating the violation would itself pose an even greater threat to the safety and health of their employees.

guard. A physical device to prevent undesired contact with a source of energy between people, equipment, materials, and the environment.

hazard. A condition or situation that exists within the working environment capable of causing an unwanted release of energy resulting in physical harm, property damage, or both.

hazard recognition. In terms of OSHA compliance, a concept based upon the premise that hazardous conditions cannot be eliminated or controlled until they are first recognized as such by the employer. An important concept since employers cannot be held in violation of a requirement if they did not recognize that the hazardous condition existed.

hearing conservation. The prevention or minimizing of occupational noise-induced hearing defects through the combined use of hearing protectors, training, the use of engineering and administrative control measures, annual audiometric testing, and the establishment of a written program.

health standard. Those standards that generally prescribe requirements for worker exposure to hazards presented by toxic substances. Such hazards usually involve the potential for long-term adverse health effects (such as those posed by exposure to lead, noise, asbestos, silica, radiation, vibration, etc.).

Hepatitis B Virus (HBV). A virus that causes inflammation of the liver. Can also occasionally be caused by toxic agents other than viral.

horizontal standard. An OSHA standard that essentially has application across a number of different industries, such as the Hazard Communication Standard and other General Industry Standards.

human error The end result of multiple factors which influence human performance in a given situation. An often over-used causal factor finding which, by itself, is not entirely descriptive of true accident cause. Human error is considered more a symptom than a cause. See *human factor.*

human factor Any one of a number of underlying circumstances or conditions which directly or indirectly effect human performance. These include physical as well as psychological factors that can potentially lead a person to make an error in judgment or action (human error) resulting in an accident.

Human Immune Deficiency Virus (HIV). Also called the human immunodeficiency virus, HIV is the virus that causes acquired immune deficiency syndrome (AIDS).

hygiene. Refers to the science of health and the preservation of well-being (named for the Greek god Hygeia).

hybrid rulemaking. A process of rulemaking that has elements of both formal and informal rulemaking procedures.

ignitable. Capable of burning or causing a fire.

ignition. The introduction of some external spark, flame, or glowing object that initiates self-sustained combustion.

illness. A condition or pronounced deviation from the normal health state; sickness. Illness can be the result of disease or injury.

Immediately Dangerous to Life and Health (IDLH). Describes the concentration of a hazardous agent that will produce irreversible health effects or death following a 30-minute exposure.

imminent danger. Refers to a work situation in which there is an immediate danger of death or serious physical harm.

Industry Rank Report. A report from OSHA's National Office in Washington, DC supplied to each local Area Office that ranks industries (such as automotive, petroleum refining, transportation, etc.) according to their lost workday injury (LWDI) rate.

incidence (or incident) rate. For OSHA record keeping purposes, the number of injuries, illnesses, or lost workdays related to a common exposure base of 100 full-time workers (working 40 hours per week, 50 weeks per year).

incident An occurrence, happening, or energy transfer that results from either positive or negative influencing events an may be classified as an accident, mishap, near-miss, or neither, depending on the level and degree of the negative or positive outcome.

industrial hygiene. The art and science of anticipating, recognizing, evaluating, and controlling occupational and environmental health hazards in the workplace and the surrounding community.

informal rulemaking: Also known as *notice and comment rulemaking,* requires OSHA provide "interested parties an opportunity to participate in the rulemaking through submission of written data, views, or arguments with or without opportunity for oral presentation." It does not require a hearing, although OSHA may hold one if it so desires. It allows the agency to look beyond any hearing records in making rules. Also, when courts review OSHA's actions under informal rulemaking, OSHA is not held to the "substantial evidence" test required under formal proceedings. Rather, the agency must only prove that their decisions and determinations are not "arbitrary" or "capricious."

injury. Physical harm or damage to a person.

Inspection Register. A registration containing the name of each establishment scheduled for inspection and the order in which these establishments will be inspected. Compiled from the establishment list and the industry rank report.

interrogatories. Part of the pre-trial discovery process. A formal set of questions, usually written, specific to the case, that must be answered by the party served, usually in writing and before the trial date.

Job Hazard Analysis (JHA) See *Job Safety Analysis.*

Job Safety Analysis (JSA) A generalized examination of the tasks associated with the performance of a given job and an evaluation of the individual hazards associated with each step required to properly complete the job. The JSA also considers the adequacy of the controls used to prevent or reduce exposure to those hazards. Usually performed by the responsible supervisor for that job and used primarily to train new employees, the JSA is also an excellent source of *paper evidence* during an accident investigation. Also known as the *Job Hazard Analysis.*

judicial branch of government. That branch of government which consists of the nation's court system, from the Supreme Court on down.

jurisdiction. Generally, the authority of the court to hear and decide a case; or, the authority of a governing or responsible body or agency to create, decide, interpret, and implement policies as in "the authority having jurisdiction."

legislative branch of government. That branch of government which consists of this nation's law-making bodies, primarily the houses of Congress.

line and staff organization In the structure of an organization, those members that are directly accountable and responsible for the daily operations of the enterprise are considered *line* management with the authority to implement or change company policy and operating procedures. Those that serve as advisors to the line and can only recommend changes are considered *staff* management.

lost workday—away from work. A day on which the employee would have worked but could not because of occupational injury or illness. This does NOT include the day of the injury or onset of illness or any days on which the employee would not have worked anyway (such as a weekend or holiday).

lost workday—restricted work activity. A day on which, because of injury or illness, the employee was assigned to another job on a temporary basis, or worked at a permanent job less than full time, or worked at a permanently assigned job but could not perform all duties normally connected with that job. This does not include the day of the injury or onset of illness or any days on which the employee would not have worked anyway (such as a weekend or holiday).

Lower Flammable Limit (LFL). Often referred to as the Lower Explosive Limit (LEL), it is the lowest concentration of gas or vapor in the air that will propagate a flame if a spark or heat source is present.

medical treatment. Includes treatment (other than first-aid) that is administered by a physician or registered professional personnel under the standing orders of a physician (e.g., nurse). This does NOT include first-aid treatment (one-time treatment and subsequent observation of minor scratches, cuts, burns, splinters, and so forth, which do not ordinarily require medical care) even though provided by a physician or registered professional personnel

mishap. An occurrence that results in some degree of injury, property damage, or both.

motion. An application to the court for a ruling or order.

near-accident. See *near-miss.*

near-miss. An occurrence or happening that had the potential to result in some degree of injury, property damage, or both, but did not. Also referred to as a *near-accident.*

noise. Commonly defined as unwanted sound and is usually expressed in decibels on the A scale (dBA), which is the scale thought to most approximate human hearing. Noise is characterized by both frequency (pitch) and pressure (intensity).

noise exposure. Exposure to any unwanted sound. Overexposure to occupational noise in the United States is considered to be 90 dBA over an 8-hour time weighted average (TWA).

Noise Reduction Rating (NRR). A measure of the effectiveness of a given hearing protector, usually expressed in decibels. Assuming a complete and perfect fit, the NRR is the difference between the sound pressure levels outside the ear and those inside the ear.

occupational injury. Any injury, such as a cut, fracture, sprain, amputation, etc., which results from a work accident or from an exposure involving a single incident in the work environment.

occupational illness. Any abnormal condition or disorder, other than one resulting from an occupational injury, caused by exposure to environmental factors associated with employment. This includes acute and chronic illnesses or diseases which may be caused by inhalation, absorption, ingestion, or direct contact.

occupational medicine. A branch of medicine dedicated to the appraisal, maintenance, restoration, and improvement of the health of workers through the scientific application of preventive medicine, emergency medical care, rehabilitation, epidemiology, and environmental medicine.

originating statute. See *Statutory mandate.*

other than serious violation (OTS). The classification of other than serious or "non-serious" violation applies when the potential consequences of the violation might be a minor illness or injury (i.e., an illness or injury which does not rise to the level of serious physical harm).

oxygen deficiency. A condition in which the concentration of oxygen by volume is insufficient to maintain normal respiration. It exists in atmospheres in which the percentage of oxygen by volume, which is 21 percent under normal conditions, drops below 19.5 percent.

permanent variance. A variance from OSHA compliance granted for an indefinite period of time. A permanent variance will be issued only if and when OSHA determines that the workplace is as safe and healthful as it would be had the employer complied with the subject standard.

Permissible Exposure Limit (PEL). An OSHA-mandated value that represents the level of air concentrations of chemical substances to which it is believed that workers may be exposed on a daily basis without suffering adverse effects. PELs are enforceable by law under the Occupational Safety and Health Act of 1970. See also *Threshold Limit Value,* or TLV.

peak sound pressure level. The maximum instantaneous level that occurs over any specified time period and is usually measured in decibels.

performance standard. An OSHA standard that essentially tells the employer *what* safety or health goal that is to be obtained. It provides details on minimally acceptable program requirements in areas such as safety training, recordkeeping, or communicating hazards to workers.

permit-required confined space. An enclosed space that is large enough and so configured that a person can enter it and perform work. This space requires written authorization prior to entry and usually has one or more of the following characteristics: (1) It has a known potential to contain a hazardous atmosphere. (2) It contains a material that has the potential for engulfing an entrant. (3) Its internal configuration presents a trapping or asphyxiating hazard. (4) It contains other recognized serious safety or health hazards.

Personal Protective Equipment (PPE). Any of a number of devices or types of equipment (hard-hats, gloves, goggles, etc.) worn to provide protection against various hazards.

plain view doctrine. Allows for the citation of employers for violative conditions that are visible to someone (including an OSHA compliance officer) who is observing from a public place.

pleadings. The statements that set forth, to the court, the claims of OSHA (the plaintiff) and the answers of the employer (the defendant).

preamble. The introductory information published with any new standard which provides great detail on the development of the final rule as well as explains the intent of OSHA in promulgating the rule. It carries no force of law.

preponderance of evidence. A point of law defined as that quantum of evidence which is sufficient to convince the judge that the facts asserted by a proponent are more likely to be true than false.

prima facia. Term of law used to describe a case that has been established without the need for further development or investigation.

principles of agency. Term of law that holds that a principal (the employer) is responsible for the acts of any of its agents (employees) that are within the scope of the agency relationship (employment).

procedural defense. Focuses on the validity of OSHA's enforcement procedures and the procedures used by the OSHRC in contested cases.

procedural mandate. Provides guidance to federal agencies on how certain decisions are to be made. Procedural mandates are basically rules governing the operation of an agency's decision making process. The most prevalent of the procedural mandates is the Administrative Procedures Act (APA) of 1946 (5 U.S.C. Section 551).

programmed inspections. Those OSHA inspections conducted as part of OSHA's regularly scheduled inspection process.

promulgate. Putting a new law or ruling into effect by making its terms known to the public.

reasonable diligence. OSHA's expectation that an employer is liable for conditions or practices which should reasonably have been known of and/or for taking preventive actions. OSHA believes employers have an affirmative duty to monitor its workplace safety and become knowledgeable of all hazards that may be present there.

repeat violation. (As defined in case law), if the same standard has been violated more than once by the same employer and there is a substantial similarity of violative elements between current and prior violations, the violation will be consider "repeated."

respondeat superior. Term of law which holds an employer responsible for the acts of employees during the course of their employment.

risk. The likelihood or possibility of hazardous consequences in terms of severity and probability of occurrence. The probability of occurrence of a loss-producing event, the chance of loss.

risk assessment. The qualitative and quantitative evaluation performed in an effort to define the risk posed to human health and/or the environment by the presence or potential presence of hazards in the work place.

risk management. The process of evaluating alternative regulatory and non-regulatory responses to risk and selecting among them. The selection process necessarily requires the consideration of legal, economic, and social factors.

safety. A measure of the degree of freedom from risk or conditions that can cause death, physical harm, or equipment or property damage.

safety professional. An individual who, by virtue of specialized knowledge, skill, and educational accomplishments, has achieved professional status in the safety field (ASSE)

Safety standard. Those standards designed to protect employees from hazards such as slips, trips and falls, lacerations and amputation from using machinery, fire hazards, and so on.

Section 6(a) standards. Also known as *1917 Base Standard,* these are the standards that OSHA adopted within the first two years after the passing of the OSHAct and are, essentially, those national consensus standards related to occupational safety and health that existed at that time. See also *consensus standard.*

Section 6(b) standards. Those standards promulgated by OSHA under the normal rulemaking procedures of the agency.

Section 6(c) standards. Also known *as emergency temporary standards (ETS).* This section authorizes the adoption of a standard without using the notice and comment rulemaking procedures. The authority to issue an ETS can be exercised any time OSHA determines that employees are being subjected to extreme danger from exposure to substances or agents known to be toxic or physically harmful and, that an emergency standard is necessary to protect employees from that danger.

serious harm. As defined by the OSHA FOM, either the permanent or temporary impairment of the body in which part of the body is rendered functionally useless or is substantially reduced in efficiency, or, illness that could shorten life or significantly reduce physical or mental efficiency.

serious violation. As defined in the Act, exists when there is a substantial probability that death or serious physical harm could result from a condition which exists, or, from one or more practices, means, methods, operations, or processes which have been adopted or are in use.

severity. In accident analysis, a measure of the degree of loss incurred, such as time away from work, resulting from an accident.

severity rate. A formula applied to OSHA-recordable injuries that result in lost time from the work place and expressed as the days lost-time x 200,000 ÷ number of hours of worked.

Short-Term Exposure Limit (STEL). The concentration to which workers may be exposed continuously for a short period of time (usually 15-minutes) without suffering adverse health effects.

silicosis. A lung disease caused by long-term exposure to silica dusts and crystals. It is marked by a loss of elasticity of the lung tissue and development of silica-containing nodules in the lungs, resulting in decreased lung function, shortness of breadth, enlargement of the heart, and often lung cancer. Also known as miner's asthma, grinder's consumption, miner's phthisis, potter's rot, and stonemason's disease.

special emphasis inspection. An OSHA inspection with focus on a particular industry (such as construction) or a particular type of work common to that industry (such as working at heights) where hazards are known to be common.

specification standard. An OSHA standard that essentially tells the employer *how* compliance is to be achieved. It may establish levels of exposure or particulars on operating certain types of equipment.

statutory mandate. A formal directive from Congress granting a particular agency the authority to act in a given area of concern, such as occupational safety and health. Also referred to as the *originating statue.*

substantive defense. Deal with the validity and applicability of a particular standard to the facts of the case, the nature of the employer's conduct, and its effect on the safety and health of employees.

System Safety Analysis. A detailed, systematic method of evaluating the risk of hazard associated with a given system, product, or program. It utilizes a variety of techniques and approaches to accurately identify, resolve, or control exposure to those hazards.

temporary variance. Essentially, permission is granted for an employer to operate under limited conditions of noncompliance with a specific standard for a limited period of time.

teratogen. A substance capable of producing physical defects in a fetus. The result may include fetal or embryonic mortality or in the birth of offspring with defects.

Threshold Limit Value (TLV) Represents the level of air concentrations of chemical substances to which it is believed that workers may be exposed on a daily basis without suffering adverse effects. TLVs are developed by the American Conference of Governmental Industrial Hygienists (ACGIH), which is a non-regulatory agency. TLVs are, therefore, not enforceable by law unless adopted by the authority having jurisdiction. *See also Permissible Exposure Limit,* or PEL.

Time-Weighted Average (TWA). An exposure averaged over a given time period, often an 8-hour work day. Threshold limit values or permissible exposure limits are often based on 8-hour time-weighted averages.

toxic. Term used to describe a chemical that has the ability to cause harmful or fatal effects upon exposure to humans, animals, or plants. The level of toxicity is normally based upon a scientific evaluation of the dose/response relationship. Basically, the evaluation of the results of varying degrees of exposure to toxic agents determines the level of their respective toxicity.

trade secret. As defined in the OSHA FOM, any confidential business device or process which gives an employer an advantage over its competitors.

unprogrammed inspections. Those OSHA inspections performed in response to particular events that occur during the inspection year, such as catastrophes, fatal accidents, and employee complaints.

unsafe act. Any act or action, either planned or unplanned, which has the potential to result in an undesired outcome or loss (injury, property damage, lost production time, etc.).

unsafe condition. Any existing or possible condition which, if allowed to continue, could result in an undesired outcome or loss (injury, property damage, lost production time, etc.).

Upper Explosive Limit (UEL). See *Upper Flammability Limit (UFL).*

Upper Flammability Limit (UFL). The concentration of a substance in air, usually expressed as a volume percent, above which combustion cannot be supported at normal room temperature because the mixture of air and fuel is too "rich" (i.e., too much fuel) and therefore, insufficient oxygen. Combustion in air of a flammable material can occur only at concentrations between the lower and upper flammability limits. This area is referred to as the *flammability range.*

vertical standard. An OSHA standard that essentially has application in only one industry, such as construction.

walk-around. An OSHA term for becoming familiar with the worksite and its employees by simply strolling through the places where employees are at work and obtaining a mental "snap-shot" view of what they are doing and the processes used by the employer.

willful violation. A violation of an OSHA requirement that is known to exist by the employer and is allowed to continue without regard to employee safety and/or health. As defined by case law, it is a violation resulting from intentional disregard or plain indifference to the Act and its regulations. Specifically, it is the employer's intent to disregard a regulation, rather than their rationalization or motive, which is relevant in determining willfulness. Employer knowledge of a standard which is being violated, rather than mere knowledge of a hazardous condition, is the essential difference between a willful violation and a serious violation. In essence, a willful violation is an intentional, deliberate, and knowing violation of the law.

witness. Any person who has first-hand knowledge of some fact related, directly or indirectly, to an issue under investigation or evaluation.

work environment. The physical location, equipment, materials processed or used, and the kinds of operations performed in the course of an employee's work, whether on or off the employer's premise, comprise the employee's work environment.

work related occurrence. If the injury or illness occurs on the employer's premises, OSHA will deem the event work related. This means that OSHA will presume the injury or illness is work related, thereby placing the burden of proving the contrary on the employer.

Bibliography

Ashford, N.A., and C. C. Caldart. 1991. *Technology, Law, and the Working Environment.* New York: Van Nostrand Reinhold.

Brauer, R. L. 1990. *Safety and Health for Engineers.* New York: Van Nostrand Reinhold.

Bronstein, D. A. 1990. *Demystifying the Law: An Introduction for Professionals.* Bocoa Raton, FL: CRC Press/Lewis Publishers.

Bronstein, D. A. 1993. *Law for the Expert Witness.* Bocoa Raton, FL: CRC Press/Lewis Publishers.

Cohen, J. M., and R. D. Peterson. 1995. *Complete Guide to OSHA Compliance.* Bocoa Raton, FL: CRC Press/Lewis Publishers.

Freeman, S. H. 1991. *Injury and Litigation Prevention, Theory and Practice.* New York: Van Nostrand Reinhold.

Institute for Applied Management and Law (IAML). 1991. *Health and Safety Law and Special Issues.* The Certificate in Environmental Health and Safety Law, Block II Course Materials. Newport Beach, CA: IAML

Johnson, Johnson, and Little. "Expertise in Trial Advocacy: Some Considerations for Inquiry into its Nature and Development." *Campbell Law Review,* Vol. 7, No. 2, Fall 1984. p.125.

Korenberg, J. P., and O. B. Dickerson. 1992. *The Workplace Walk-Through.* Bocoa Raton, FL: CRC Press/Lewis Publishers.

Lisella, F. S. 1994. *The VNR Dictionary of Environmental Health and Safety.* New York: Van Nostrand Reinhold.

Madden, M. S. 1992. *Toxic Torts Deskbook.* Bocoa Raton, FL: CRC Press/Lewis Publishers.

Matson, J. V. 1994. *Effective Expert Witnessing.* 2nd Edition. Bocoa Raton, FL: CRC Press/Lewis Publishers.

Michaud, P. A. 1995. *Accident Prevention and OSHA Compliance.* Bocoa Raton, FL: CRC Press/Lewis Publishers.

National Safety Council. 1988. *Accident Prevention Manual for Industrial Operations, 9th Edition.* Chicago: National Safety Council.

Nwaelele, D. 1994. *Health and Safety Risk Management: Guide for Designing an Effective Program.* Rockville, MD: Government Institutes.

Public Law 91-596. *The Occupational Safety and Health Act of 1970.* 29 December 1970. As amended by Public Law 101-552 (section 3101), 5 November 1990. 91st Congress. Washington, DC: U.S. Government Printing Office.

Railton, W. S. 1992. *OSHA Compliance Handbook.* Rockville, MD: Government Institutes.

Spencer, J. W. 1992. *Health and Safety Audits.* Rockville, MD: Government Institutes.

U.S. Department of Labor. 1994. *Regulations Related to Labor-Construction Industry.* Occupational Safety and Health Administration, Code of Federal Regulations, Title 29, Part 1926. Washington, DC: Office of the Federal Register, National Archives and Records Administration.

U.S. Department of Labor. July 1994. *Regulations Related to Labor-General Industry.* Occupational Safety and Health Administration, Code of Federal Regulations, Title 29, Part 1910. Washington, DC: Office of the Federal Register, National Archives and Records Administration.

U.S. Department of Labor. 1977. *Investigating Accidents in the Workplace: A Manual for Compliance Safety and Health Officers.* Occupational Safety and Health Administration, OSHA Publication Number 2288. Washington, DC: U.S. Government Printing Office

U.S. Department of Labor. 1994. *OSHA Field Operations Manual.* Occupational Safety and Health Administration, Washington, DC: U.S. Government Printing Office

U.S. Department of Labor. 1993. *OSHA Technical Manual.* Occupational Safety and Health Administration, Washington, DC: U.S. Government Printing Office.

Vincoli, J. W. 1994. *Basic Guide to Accident Investigation and Loss Control.* New York: Van Nostrand Reinhold.

Wolff, K. 1995. *Understanding Workers Compensation: A Guide for Safety and Health Professionals.* Rockville, MD: Government Institutes.

Index

F

G

H

I

L

M

N

O

P

R

More Safety and Health Titles

Fundamentals of Occupational Safety and Health

This book covers the basics safety and health professionals need to control hazards, prevent losses, and protect the health and lives of workers. This book balances the management of safety with relevant science and practical aspects of complying with regulations. A basic text and desk reference in one volume, **Fundamentals of Occupational Safety and Health** is a book professionals will refer to repeatedly throughout their careers.

Softcover, Index, 429 pages, 1996, ISBN:0-86587-539-1 **$49**

Safety Made Easy: A Checklist Approach to OSHA Compliance

This book provides a simpler way of understanding your requirements under the complex maze of OSHA's safety and health regulations. Instead of explaining the OSHA CFR in a chapter and verse fashion, the authors have created checklists for OSHA compliance organized alphabetically by topic. Each checklist begins with a brief description of the objectives of the listed items, followed by the required actions and corresponding standards, and where appropriate, training, personal protective equipment, and recordkeeping requirements.

Softcover, Index, 192 pages, 1995, ISBN:0-86587-463-8 **$49**

Total Quality for Safety and Health Professionals

This book will show you how to apply the successful concepts of Total Quality Management to your Industrial Health and Safety program. Author F. David Pierce explains what Total Quality is and how it can be used to improve the safety process. Using his own experiences and numerous case examples, Pierce examines both the fundamentals and implementation of Total Quality Management. In addition, he discusses common roadblocks to Total Quality and how you can overcome them.

Hardcover, Index, 229 pages, 1995, ISBN: 0-86587-462-X **$59**

"So You're the Safety Director!" An Introduction to Loss Control and Safety Management

Let author Michael V. Manning's narrative approach and easy-to-follow writing style make it seem like you've hired him to help you start—or upgrade—your safety program, which is exactly what hundreds of companies have done. Manning walks you through the do's and dont's of establishing and evaluating your company's safety program.

Softcover, Index, 186 pages, 1995, ISBN 0-86587-481-6 **$49**